SNEAKERS
UNBOXED
STUDIO TO STREET

Edited by Alex Powis

the DESIGN MUSEUM

CONVERSATIONS ON
SNEAKER DESIGN WITH

TOM ASTRELLA
FRANCK BOISTEL
JACQUES CHASSAING
MARINA CHEDEL
BEN COTTRELL
MATTHEW DAINTY
PETER FOGG
RYAN FORSYTH
JOE FOSTER
NIC GALWAY
ROMAIN GIRARD
BENJAMIN GRENET
SAM HANDY
ASHA HARPER
CHRIS HILL
STEPHANIE HOWARD
TILL JAGLA
SARA JARAMILLO
MURIEL JUNG

JEAN KHALIFÉ

HELEN KIRKUM

JEAN-PHILIPPE LALONDE

CHRIS LAW

TUAN LE

CHARLOTTE LEE

STEVE MCDONALD

CARLY MCKENZIE

NICOLE MCLAUGHLIN

PETER MOORE

ANDREA NIETO

SAMUEL PEARCE

RIAN POZZEBON

SUSI PROUDMAN

DAVID RAYSSE

SAMUEL ROSS

JULIANA SABAT

KIRSTEN SCHAMBRA

CHRIS SEVERN

STEVEN SMITH

ALEXANDER TAYLOR

DANIEL TAYLOR

DIRECTOR'S FOREWORD

As cultural artefacts go, sneakers offer an extraordinary insight into the cultural, commercial and industrial life of a society. A single pair can be the thread connecting a particular sport, music scene or subculture with a piece of cutting-edge manufacturing technology. But in a world that moves as fast as the sneaker industry does – a world that is always thinking about the next iteration, the next drop – there is not always time for reflection. Indeed, many of the sneaker brands themselves would acknowledge that they are so focused on the future that they have sometimes neglected to document, or even store, their past.

This is where museums, and a museum of design in particular, become important. *Sneakers Unboxed: Studio to Street* offers historical perspective, but it is far from a mere history. The exhibition, and this accompanying book, strive to represent where we are today, including up-to-the-minute technical innovations. In fact, this was our original motivation for staging the exhibition. The pace of innovation in the sneaker industry is so rapid, and the role of design so instrumental, that it makes a fascinating relationship for the Design Museum to explore. The number of limited-edition sneakers released every year has accelerated dramatically, and the trend cycles are now so fast – a matter of months – that traditional design production timelines simply can't keep up. Robotic weaving, 3D printing, customisation, crowd-sourced design – all of these processes are revolutionising the sneaker production model.

It used to be that all of that originality and inventiveness trickled down from secretive design and innovation labs to athletics tracks and basketball courts. But increasingly, innovation is also being driven from the street, by subcultures focused on grime or hip-hop, or by avid communities of sneakerheads. Style, in its almost infinite permutations, is now perhaps an even greater generator of sneaker culture than performance. Of course, all this boundless creativity and production has an environmental cost, and the sneaker industry is still in the early stages of addressing it. Design will play a critical role.

All these issues are explored in the exhibition, and in the various and compelling interviews with the leading sneaker designers that are collected in this book. The Design Museum would like to thank StockX in particular for supporting the exhibition, and for helping us celebrate the cultural significance and complex social history of these seemingly everyday objects.

Tim Marlow
Chief Executive and Director of
the Design Museum

CURATORIAL FOREWORD

More than one billion pairs of sneakers were sold in 2020 alone, and new models appear on websites and blogs virtually every day. The sneaker is probably the most ubiquitous designed object, worn as everyday footwear, as performance sportswear, as an aspirational style item, or perhaps not even worn at all – instead bought expressly to be kept, and boxed up, as a rare collector's item. Once seen as only being in the purview of athletes, amateur or professional, sneakers are now worn on all occasions: at work, in clubs and bars, and even to weddings.

The origins of the popularisation of sneakers worn for style can be found in the late 1970s, across a number of youth cultures in the US and the UK. Thanks largely to young people, from diverse and often deprived inner-city neighbourhoods, who sought affirmation through the way they dressed and which sneakers they wore, a style that was formerly only associated with sports took on a different meaning – that of role and status. The codes established therein – exclusivity and uniqueness, newness and freshness – are what continue to drive sneaker culture today.

The exhibition *Sneakers Unboxed: Studio to Street*, which this book accompanies, aims to unravel some of that history behind the sneaker phenomenon. It focuses on the people who popularised sneakers as highly covetable style items, as well as the designers and engineers behind some of the key innovations that have helped to break records and improve performance.

The exhibition is divided into two sections: Style and Performance. The first section focuses on notable figures in the history of sneaker design and culture, including basketball stars from the 1970s, such as Dr J, Kareem Abdul-Jabbar and Walt 'Clyde' Frazier, who played on streetball courts across the US. Their influence was particularly palpable at the Rucker Park tournament in Harlem, New York. Huge crowds attended, dressed to impress, often wearing the same sneakers as their heroes on the court. In the 1980s, musicians such as Run-DMC and, of course, basketball superstar Michael Jordan made sneakers as style a global phenomenon.

The exhibition also charts the influence of specific local youth cultures, such as the UK football casuals, the South African bubble heads, the Mexican Cholombianos and the international skateboarding community. What they all share is a desire to express identity through their choice of sneaker. In doing so, they have inadvertently galvanised the exponential growth of the industry.

Collaborations between sports brands, small retail and fashion designers, now commonplace, have become the site of an experimental design language freed from the needs of technical performance. The earliest limited editions such as New Balance's long-standing collaboration with sneaker boutique Mita and now defunct streetwear brand Mad Hectic, as well as sports brand collaborations with fashion designers such as Y3, emerged from a relatively small area around Harajuku and Aoyama in the late 1990s. Leveraging a small, dedicated following, by creating limited editions with independent skate and sneaker stores across the world, the way sneakers were bought and sold was changed forever. Queues outside sneaker stores became a regular sight. Buying to sell became the norm and the resulting resale market is now an industry in its own right, with online platforms, such as StockX, recording sky-high prices, and even offering platform-specific releases of sneaker models.

Beyond the stereotypical image of 'sneakerheads' obsessively queuing for days for a limited-edition pair, and then reselling them online, lies a much longer history of engagement with the design history of sneakers. Sneaker collectors have acted both as the gatekeepers of sneaker culture and the historians of sports shoes. Driven by a desire to own all versions of a specific model or super limited editions, or by an obsession with a specific brand, collectors have played a key role in the development and documentation of the sneaker industry.

The Performance section of the exhibition uncovers the history of design and innovation. The design process behind key sneaker models that are considered innovative and ground-breaking continues to revolve around recurring design concerns. Traction, cushioning, fit, stability and energy return are still seen as central to improving athletic performance. However, different types of athletic movement demand different footwear design considerations: basketball players may need heightened traction and stability on court, which has resulted in now-iconic models such as the Jordan line, Converse Weapon or Nike Foamposite; endurance runners, on the other hand, require maximum comfort while conserving as much energy as possible. Designing to strike the balance between cushioning and firmness has resulted in some of the most innovative running shoes, including the ASICS Gel-Lyte III and adidas Boost. Some claim up to five per cent of energy return, which has resulted in several records being broken. One example is long-distance runner Eliud Kipchoge, who made history in 2019 with a prototype of the Nike Alphafly NEXT% by running a marathon in under two hours. Results like this ignite a recurring debate – famously played out at the 1968 Olympics, with the banning of the PUMA Tahoe for having too many small spikes – around where sports equipment ends, and technology doping begins.

This section also considers the most pressing design concerns of our time, and one that the industry and designers are more and more preoccupied with: sustainability. This particularly includes circular design and more ethical manufacturing practices. Innovation in materials such as Mylo, a leather made from mushrooms, as well as experimental manufacturing techniques such as adidas' FUTURECRAFT.STRUNG, a 3D knitting robot, and increased transparency in supply chains pioneered by smaller brands such as Veja and Satoshi Studio – these are just some of the innovations being made towards reducing the impact of sneaker production on the planet.

Designers such as Helen Kirkum, and customisers such as Japanese company Recouture and Leeds-based Adikoggz, are creating more environmentally conscious practices by extending the lifespan of sneakers through upcycling, remaking and refurbishing.

The history of the sneaker is a reflection of cultural, societal and technological change. By looking at, recording and analysing this history, much can be anticipated of the future of the industry. Many of the people who made sneakers what they are today are largely missing from the industry now. A focus on more inclusivity in education and diversity in the industry as a whole, coupled with more collaborative design and co-creation processes may just ensure that the legacy of those who created sneaker culture is acknowledged, making the future of sneaker design a little bit brighter.

Ligaya Salazar
Exhibition Curator

INTRODUCTION

Footwear has long been attached to societal constructs and cultural movements, ever-shifting in its coded meaning and relevance to humanity. At its most basic, footwear is a functional item built to serve a purpose. But place it within society and it becomes a symbol of cultural tribalism, outwardly telling the world what you do or do not represent. Sneakers in particular have condensed these tropes – and our relationship with footwear – and put them into hyperdrive. Operating for decades at the cutting edge of technology, sports performance and culture, the sneaker industry acts as an innovator and a facilitator for human achievement and progression. Whether it's enabling us to run faster, to manufacture in ways we hadn't imagined, or, more recently, to consume like never before.

Like any good sneaker design, this book has been created with a functional purpose. The world does not require more books rehashing the same stories of iconic sneakers from the past, nor does it need one that overly aggrandises a small list of familiar sneaker designers. We've experienced that enough in the past, which has in turn contributed to creating a skewed view of an industry built by a few heroic pioneers. Like most things in life, the truth is more complex. Those pioneers, as visionary as they were, did not act alone. The success of any sneaker at any moment in time is thanks to a multitude of often unnamed team members and collaborators, engineers and manufacturers. This book reveals the collaborative effort of the sneaker industry in depth, for the first time. No person is an island; no sneaker is created solely by its designer.

Within these pages, you will find thoughts, opinions, stories and insights from more than forty people involved in making sneakers over the last sixty years – from those behind the adidas Superstar and Nike Air Jordan 1, to YEEZY and Nike's ISPA initiative. They were all asked the same questions and given the freedom to answer whichever they saw fit. This has resulted in a book built from a multitude of voices, backgrounds, experiences and opinions, offering an honest – and at times cutting – insight into the process of creating sneakers. It addresses how that process has evolved, and how it needs to change going forward. You'll find people unknowingly agreeing with each other and sometimes directly contradicting each other. It is this freedom of expression and diversity of opinions that give the book its true value. The same value that is an essential ingredient in any successful design team, as is revealed by the contributors themselves.

Unlike a traditional exhibition catalogue, this book isn't a mirrored reflection of the *Sneakers Unboxed* exhibition; it offers instead a continuation of the conversation started by the exhibition, and provides an alternate perspective on the same story. It takes a look inside the minds on those who dreamed into existence the sneakers featured in the exhibition, and at the same time adds something new to a conversation that at times can feel repetitive and rehearsed.

Most importantly of all, this book is a functional object created for you, our reader. Use it however you wish. Read it from cover to cover, or jump between themes that capture your interest. Use the index at the back to navigate your way through the text from the perspective of individual voices, or ignore the names and just glean the wisdom. Be inspired by it, disagree with it, learn from it and challenge it. However you approach the book, digest the information provided and use it as a seed of inspiration for your own story. This book is nothing if not a starting block for conversation and thought. Highlight it, bend the corners of the pages, scribble your thoughts in the white spaces. Make it yours. Like any good pair of sneakers, it is here to be used and loved.

Alex Powis
Editor

IN MEMORIAM This book is dedicated to the memory of my friend and guide, Gary Warnett. I hope he would have enjoyed it.

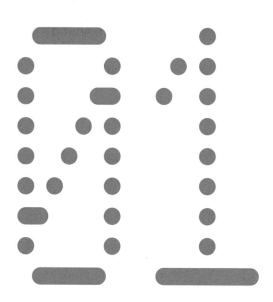

STYLE AND CULTURE

STYLE AND CULTURE

Looking around us today, it's hard to imagine a time when sneakers, style and culture weren't intertwined. They appear symbiotic, with the boundaries between them too blurred to define. What influences what? Who takes the lead from whom?

Rewind to 1984 and a twenty-one-year-old Michael Jordan signed with an even younger Nike, collectively putting into action a partnership that would rewrite the rules of athlete endorsements, culture, influence and consumerism on an unfathomable scale. Outside the world of sport, Run-DMC inked a deal with adidas in 1986 that ushered in the era of endorsement deals between musicians and sports brands. Two distinct moments; the world was changed forever.

Bold innovation in the 1980s was often matched with equally impactful colours. The adidas ZX '000 series took this to an extreme in 1988. Its vivid colourways, chosen to attract runners' attention, also caught the eyes of acid house ravers, and the ZX '000 series was subsequently adopted as a symbol of the movement. Whether pounding pavements or the floors of disused warehouses, the technology performed. Was this adoption simply due to those colours, or was it the comfort, innovation and prestigious price tag that appealed? It's hard to pin down, and the nostalgic eye of hindsight doesn't help with clarity.

Memorable leaps in innovation often inadvertently influence style and culture, continuing to do so through the generations of designers who follow. In a poetic turn of events, this book features influential designers who unknowingly cite other contributors' work as influences on their own careers. Chris Severn's Superstar, Peter Moore's Air Jordan 1 and Steven Smith's Instapump Fury are repeatedly name-dropped by designers who themselves have gone on to shape the sneaker industry and culture.

Why do some designs transition from sports to cultural icons? What enables a sneaker to be timeless? And how much input, if any, does the designer have in all of this? Some sneakers appear impervious to ageing, moving from decade to decade as if time doesn't exist. But there are plenty that made waves on their arrival but which fail to gain traction today, possibly due to dated technology or a lack of cultural resonance, or perhaps simply because of a lack of investment in bringing them back.

There is no simple answer to what defines culture, style or popularity. All we can do is explore and question in the hope to better understand.

ICONS

STEVEN SMITH The New Balance Trackster, to me, was the king of it all ... the first modern sneaker with a midsole.

JACQUES CHASSAING The most important sneaker designs to date are the adidas Stan Smith or Superstar, Nike Air Max or Air Jordan, and Nike Air Force 1. These iconic shoes shaped and are still shaping your style today.

PETER MOORE The Nike Air Jordan 1 was my first design, and for the most part my only design from concept to finish.

STEPHANIE HOWARD The Nike Air Jordan 1 brought the entire dynamic of product endorsement to a new level.

STEVEN SMITH The Superstar is one of the all-time greats. It was the first fusion of sport and lifestyle.

NIC GALWAY The Stan Smith and the Superstar are important because they represent the dialogue between sport and culture.

ABOVE   Nike Air Jordan 1, Who said man was not meant to fly,
advertisement, 1985

ICONS

ABOVE Stan Smith, London, 2018. © Juergen Teller, All Rights Reserved
OPPOSITE Reebok Instapump Fury, first released in 1994

CHRIS SEVERN We almost made the Superstar into a tennis shoe. We wanted it to look different, so we gave it a different sole profile, and that became the Stan Smith.

JOE FOSTER Reebok introduced soft, tanned leather, which changed the feel and comfort of sports shoes.

TOM ASTRELLA My favourite innovation was conceived twenty-five years before it was physically possible: power laces!

STEVEN SMITH The Nike Air Sock Racer made me want to explore the future. It led me to create the Reebok Instapump Fury.

SAMUEL PEARCE The Reebok Instapump Fury made me dream of a future with flying cars and life on other planets. I spent a year trying to persuade my parents they would be a 'useful' addition to my footwear rotation.

ICONS

DANIEL TAYLOR The archetype that a sneaker should have laces is still a barrier that pushes designers to innovate. In this respect, the PUMA Disc System was decades ahead of its time.

SAM HANDY I can remember the first time I saw an Air Max 90 and a ZX 8000. The kids in school had no idea what those shoes really did.

MURIEL JUNG The original adidas ZX and Equipment concept series, or the Ultraboost, all influenced the evolution of running footwear.

TILL JAGLA The first time a brand challenged the sole unit's construction was in 1988 when adidas invented Torsion, the first mid-foot bridge, allowing for a more flexible running shoe.

JACQUES CHASSAING With the ZX '000 series for Torsion, we said, 'If you have a new innovation, you have to make an impact.' Not only function, but also material and colour.

Labels in sketch: FORE FOOT STRAP / TOE OVERLAY / PU AIR CASSETTE / MORE OPEN LATERAL side / TERRA TRAIL / FGG

ASHA HARPER I love the attitude and vibe the Nike Air Max Plus emits. It's sporty and fierce. At the same time, it's sleek and sophisticated. It challenged old ways and got TPU welding to work in ways it had never worked before.

PETER FOGG In 1997 my design for the Nike Air Terra Humara became a cultural icon – it was a big hit with famous actors and models in LA and New York. I was not trying to make an icon. I was trying to make the first trail shoe with Vis Air.

STEPHANIE HOWARD The reissue in 1997 of the 576 started the retro wave for New Balance and it spread to other brands. That retro category has never officially left the sneaker ecosystem.

OPPOSITE   adidas Equipment Running Support, 1991
ABOVE   Peter Fogg, design sketch of Nike Air Terra Humara, 1996

KIRSTEN SCHAMBRA A past innovation that I love is the Nike Air Rift. Designed to be minimal and work with the mechanics of the foot, it was the most comfortable and strange-looking shoe.

NIC GALWAY The season I designed the Y-3 Qasa, I had a conversation with Yohji Yamamoto, and he told me he was bored and that all sneakers had started to look the same. He said, 'Show me something different.' It gave me permission to create.

CHRIS HILL I was influenced by the early Nike SB era, Reebok × ICECREAM, and Reebok pop culture-themed shoes of the early 2000s.

JEAN KHALIFÉ The first pair of sneakers that I absolutely adored and made me want to work in this industry was the adidas Gonz Pro 2. They were so different – over-designed and over-moulded.

DANIEL TAYLOR It was the Mihara Yasuhiro and Alexander McQueen collaborations with PUMA that set the standard for me. The creativity of those products completely blew my mind.

TUAN LE AND1 products had to have that flavour of hip-hop music and trash talk, or nobody would care. It was music in motion when a player ran down the court and dunked the ball in the AND1 Tai Chi.

BENJAMIN GRENET I designed the Salomon XT-6 without any quest to be cool. I designed more by removing things than adding style to the shoe.

ABOVE   Salomon Advanced SS21 Collection, 2021
OPPOSITE   adidas, YEEZY 350 Glow, 2019

STEPHANIE HOWARD Nike Flyknit technology changed the way all footwear companies thought about production, and the potential to radically revolutionise it.

JEAN KHALIFÉ Boost technology redefined the meaning of comfort and responsiveness.

SAM HANDY Ultraboost was not built to be a fashion shoe at any level. It became a fashion shoe because it was premium, exciting technology.

DAVID RAYSSE Kanye turned our industry on its head. The significance of the YEEZY 350 and Kanye's impact is still reverberating. The culture is bigger than sport now.

MARINA CHEDEL Nike Go FlyEase changed the game by focusing on utilitarian innovation.

CULTURE

PETER MOORE A designer who sits down to design a cultural success will fail, no matter who that designer is.

ANDREA NIETO Culture influences footwear design, and design influences culture.

NIC GALWAY We are constantly moving forward in cycles, building the narrative of what sneaker culture is, both positive and negative. It's a constant dialogue that collectively adjusts and reinvents itself.

DANIEL TAYLOR Identifying cultural shifts is a way to connect directly with people through design.

HELEN KIRKUM I think it is really important to consider how a shoe will be perceived in the wider world when designing it, but also not to let it weigh you down.

DAVID RAYSSE I never think about cultural impact when designing. I am at the mercy of the design.

CULTURE

ABOVE   Mike Skinner, Lower Eastside, New York, 2002. Skinner is wearing
the Nike Air Max Plus

JACQUES CHASSAING When I designed shoes like the adidas ZX or Forum, I was just thinking about performance. Cultural expression was something I never thought about.

CHRIS HILL Cultural impact isn't something I ever think about when designing. However, I do think about whether it will pay authentic homage to the culture or help push the culture forward.

STEPHANIE HOWARD The more innovative the solution, the higher the likelihood of it becoming a cultural success, because it has challenged norms.

SAM HANDY Your top-end fashion consumer is drawn to a sports brand for its technical performance innovation. They're not drawn to a sports brand for fashion. The sports brand's contribution to fashion is technology and innovation. That's where you drive the industry, and you drive trends.

CHRIS LAW If you are trying to replicate or inform culture, you're probably trying too hard.

STEPHANIE HOWARD There are a few key reasons why some sneakers become a significant force, and it often comes down to how they changed the status quo.

HELEN KIRKUM When sneakers are linked to cultural moments, athletes and celebrities, or even groups of people, mindsets, beliefs or ways of life, they can become an accessible vehicle to express your point of view.

CULTURE

TILL JAGLA In many cases, it wasn't the design or purpose itself, but rather the context – a movement, a famous athlete, or even the price – that turned sneakers into street currency. Youth has always looked out for statement pieces and identifying features to represent a certain subculture.

DAVID RAYSSE When Run-DMC immortalised the adidas Superstar, the worlds of athletic footwear and popular culture were forever intertwined.

STEVEN SMITH Kanye has elevated sneakers to precious collector items. He has carried that torch from Michael Jordan over from sport into lifestyle.

SAM HANDY Externally, it could have looked like a marketing ploy, that we put Kanye in triple white Ultraboost, to make them cool. But it actually surprised everyone. We couldn't keep up with demand. That's one of the reasons they were so rare.

CHARLOTTE LEE Culture impacts design massively. Whether it's diversity and inclusion, sport or fashion – it all affects how we design, and the product we create.

TOM ASTRELLA Culture and fashion dominate top-tier sneaker design today. The most coveted sneakers are no longer the best performance sports shoe. Designs have evolved into pieces of art and expressions of culture rather than solely performance equipment.

ABOVE Tyshawn Jones wearing adidas Tyshawn, kick-flipping over a cone, New York, 2019

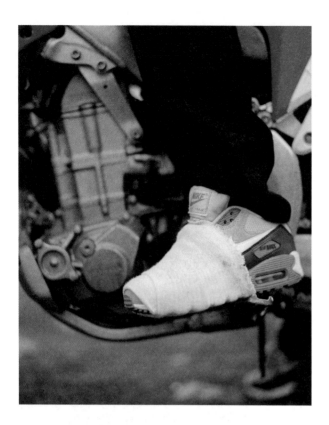

JULIANA SAGAT A sneaker can transcend differences between people. It pulls them together, creating new community and unity.

ASHA HARPER Sneakers are uniform for the majority of cultural identities. Identifying with a particular culture gives people feelings of belonging and security.

BEN COTTRELL An accessible garment such as footwear creates an instant dress code among groups – a reliable identifier to recruit or deter others.

CHRIS LAW It's a designer's choice over how much they want to serve existing culture or challenge it.

MATTHEW DAINTY When designing footwear, a designer cannot predict future cultural impact.

SAM HANDY The individual designer doesn't have a lot of influence on whether a shoe gets commercially accepted. If you build an amazing performance shoe, it's automatically cool to a certain audience.

TUAN LE I can create a perfect shoe, and whether it will become a cultural icon, I could never know.

BENJAMIN GRENET Some of my designs have become popular in the street ten years after their creation. It's a strange feeling. The sneakers market likes new things, but not unproven things.

OPPOSITE Cian Oba-Smith, *Untitled*, from the series *Bikelife*, 2015
ABOVE Tinker Hatfield, design sketch of Nike Air Max 90, 1988

CULTURE

ABOVE   Owen Nieder wearing Vans Style 38 (renamed the Sk8-Hi in 1994),
Sanoland, Cardiff-by-the-Sea, California, 1984

CULTURE

RIAN POZZEBON If the design team is paying attention to what's going on, but also has an honest story to tell, it'll resonate within the market.

SARA JARAMILLO A lot of brands are hyper-aware of what's going to happen, what's going to be the next trend or the next success. But I think we as consumers decide if that's a yes or a no.

SAMUEL ROSS There's so much surrounding collateral to having a culturally successful shoe. A lot of it is down to celebrity, ideation and concept, heritage and history, and who holds the key for soft power on a regional or national stage. All of those elements need to interlink and overlap for you to have a cultural hit. A designer is only a quarter of that conversation, at max.

PETER MOORE Cultural icons are created by movers and shakers and consumers – they are not designed.

NIC GALWAY Ultimately the consumer is the one who will take a product and give it a personality beyond the brand.

ABOVE Julien Boudet, GT3 RS, *Tailwind et Shox*, 2020, archival inkjet print on paper (unframed) 59 × 39 ⅜ in (150 × 100 cm), JB-032.3, edition of 3 + 2 AP, courtesy of the artist and Stems Gallery

CULTURE

TIMELESSNESS

SARA JARAMILLO It's really hard to say: 'This is going to be a timeless design.' We all hope that happens, but that's not entirely in our control. That job goes to the consumer. If they believe in it, then it's going to be a timeless design, or at least a success.

MURIEL JUNG The real timeless designs were created to fulfil a purpose, not to feed into a brief fashionable need.

SAM HANDY When you look through the products that have endured, they're the ones connected to performance innovation and sport. That's what fuels the industry.

CHARLOTTE LEE Often the sneakers that are visually different or pioneer a new technology are the ones that live on beyond the trend associated with the design.

DAVID RAYSSE The sneakers that transcend generations are pure on some level. They are distilled down to a minimal number of components, thus immune to 'style' and its ever-changing, fickle masters.

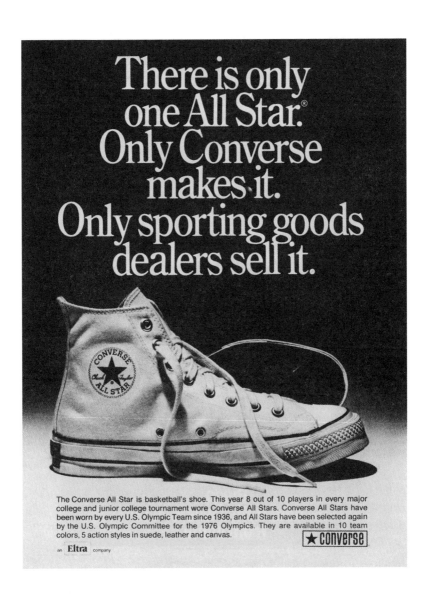

The Converse All Star is basketball's shoe. This year 8 out of 10 players in every major college and junior college tournament wore Converse All Stars. Converse All Stars have been worn by every U.S. Olympic Team since 1936, and All Stars have been selected again by the U.S. Olympic Committee for the 1976 Olympics. They are available in 10 team colors, 5 action styles in suede, leather and canvas.

an **Eltra** company

★ **converse**

ABOVE   Converse, There is only one All Star, advertisement, 1972

TILL JAGLA If a product only features one or two iconic design markers, it can become a timeless icon. Every successful shoe has these franchise identities: the Stan Smith carries the heel tab; Superstar the shelltoe; NMD the plugs; ZX the heel cage.

CHRIS LAW Purity – and doing as little design as possible – but retaining character and flavour, are key success factors for timeless designs.

MURIEL JUNG The key to timeless design is to cut out the influence of style trends in your decision-making chain as far as possible. Sometimes this is harder than it sounds.

CHRIS SEVERN If a shoe gets a second life or a third life, so be it, but my original task wasn't to create fashion footwear.

JEAN KHALIFÉ In general, the simpler you get in terms of design language, process and natural materials, the more timeless you get.

STEVEN SMITH If you do your job as a designer and create an object of desire that solves problems for the users, they become memorable.

CHRIS HILL I don't think there is any answer or formula to creating a timeless design. It starts off with a great design, then the consumers and cultures decide the rest.

SAM HANDY Credibility is essential to building long-term design icons. The thing that unites

NMD, YEEZY 350 and Ultraboost is the credibility of purpose around Boost. Boost was a whole new experience for the industry.

JEAN KHALIFÉ White vulcanised tennis shoes from the 1960s are often seen as the most timeless and effortless footwear. That's because of the shape, the simple construction and how the materials age over time.

DANIEL TAYLOR When a sneaker transcends years or even decades, it becomes embedded in culture itself. The emotions and memories it evokes in us elevate it to iconic status.

CHRIS SEVERN If a shoe lasts as a big seller for five to ten years, that's a smashing success. If it lasts for fifty years, as has the Superstar, that's incredible.

STEVEN SMITH Some styles become trapped in a moment by colour or purpose. Others transcend to the next level of timelessness at introduction. Sometimes it takes time for people to discover a future classic.

CHRIS SEVERN Stan Smith lives on because the shoe has a very long life, and his name is on that shoe.

CHANGE

TUAN LE In the beginning, there was no title for this job: my first business card titled me as a 'graphic designer'.

CHRIS SEVERN Times are changed. When I was immersed in this industry, there weren't that many people like me. Now there are lots of designers, and lots of collectors. This whole sneaker thing is a phenomenon that has occurred in a major way within the last two decades.

STEVEN SMITH When I started, this wasn't really an accepted career. What we did made it interesting and appealing for the next generation. In the early days, we had to know all aspects of shoe-making, from tanneries to pattern-making and sewing, to mould construction, and we did all the work by hand. There is something special about the craft aspect that is lost today with the purely digital approach and compartmentalised role of the designer.

CHRIS LAW The role of a designer today differs from those that paved the way. I don't think there will be another Tinker Hatfield, Jacques Chassaing,

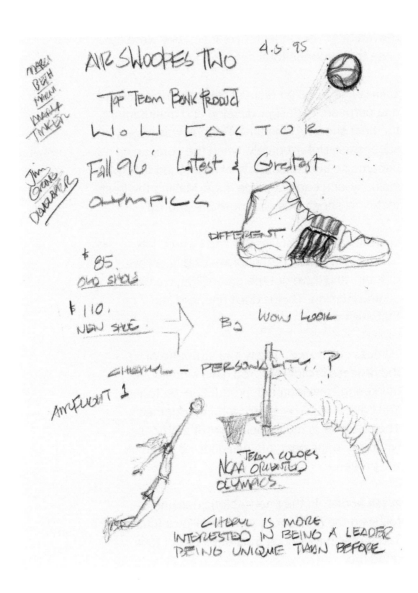

Steven Smith or Christian Tresser. They were the ones that brought 'total new' to the table.

TUAN LE The design team at Reebok originally had one purpose: to design, develop and manufacture the best shoes. At the time, Reebok was hugely successful. Unfortunately, when the company became so successful, lots of money led to lots of jealousy, and egos became huge. Making the best technical shoes was no longer the priority.

CHRIS SEVERN Horst Dassler and I would talk, I'd share my ideas, he'd approve or disapprove of them, and if we got the green light, we were up and running. Things don't happen like that anymore.

JACQUES CHASSAING Today, the structure of the process is different. You have a designer, marketing, development, production, factories. You have a lot of partners working together. The designer is much more part of a process, whereas, in the past, the designer was leading the process.

PETER MOORE In the past, athletic designers followed the philosophy of form follows function. Today, this may still be true, but it better look good too.

HELEN KIRKUM A designer's role today can become very layered and complex. The design of a sneaker is no longer solely about the design lines, functionality and durability, but also the social and political context of the sneaker.

TILL JAGLA Today, briefs are based on consumer insights and market needs. Organisations need to create space for creativity. If every product needs to be backed up by stats, facts and figures, brands can't develop in order to excite and surprise people.

SARA JARAMILLO YEEZY felt like a start-up at the beginning. We had freedom. Not having a marketing team telling you what to do because of sales or something not being on-trend – heaven for designers.

SAM HANDY As the sportswear industry explodes, the pressure on individual products to succeed gets higher and higher.

ABOVE   David Sims, Launch, 2020, for Aries × New Balance

CHANGE

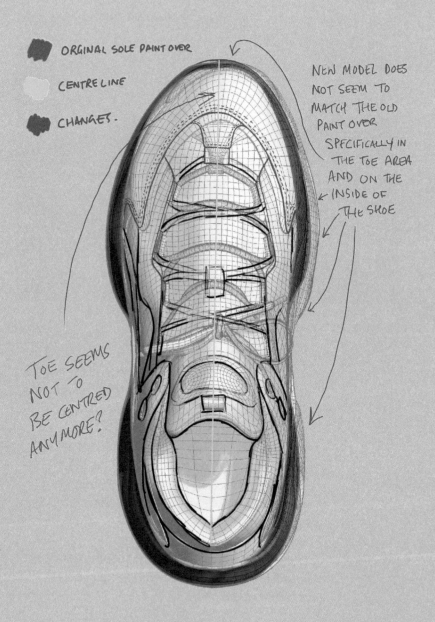

ABOVE   Tom Astrella, design sketch of Project-X with correction notes, 2019.
The initial model built in Modo (3D) and corrections sketched over in Procreate

CHRIS HILL Designers of the past really focused on great design and innovation. You still have that now, but now there are so many more things to consider that play no part in the shoe's function, or that really bring any value to it.

ASHA HARPER Sneaker design has always been based on performance. In recent years the shift has been towards lifestyle, by adapting the definition of what it means to be an everyday athlete.

JEAN KHALIFÉ With the industry changing, the designer is more and more focused on cultural insights, mixed with sport. With this shift in the market, a modern designer needs to develop marketing skills as well.

MURIEL JUNG There are benefits to being a designer nowadays. We have way more creative tools and technical resources. Freehand sketching can be essential, but it is a privilege to choose your favourite and most efficient tool, from paper to iPad to VR goggles.

TOM ASTRELLA Today, footwear designers have a whole arsenal of tools at their disposal, and those have evolved greatly over the last decade. Pen and paper have given way to iPads; 2D is currently giving way to 3D.

BENJAMIN GRENET The role of the designer has changed totally. Fifteen years ago, when I designed the Salomon XT-Wings, everything was possible in terms of materials and workload, because Asian manufacturing was so cheap at that time.

We put so much stuff on the same shoe. It's totally different now.

STEPHANIE HOWARD New innovations in production have created more avenues for design concepts, so the early ideation stage is more robust than it used to be.

DAVID RAYSSE Sneakers from twenty years ago feel like tanks compared to what's available today. Knitting, woven jacquards, high-frequency welding, 3D printing, polymer technology in foams and cushioning technologies – it's just incredible.

DANIEL TAYLOR We have entered a 'new democracy' in sneaker design. Everything, and everyone, is a potential influence on what we do as designers.

HELEN KIRKUM Our access to our audience and wearers is much more direct and casual than it ever used to be. Today, many designers can become acknowledged in their own right for their work, rather than being lost inside large companies.

DAVID RAYSSE Social media has elevated the stature of designers to a level of celebrity. In the past, only designers who were the heads of fashion labels or their own labels enjoyed this level of notoriety. Now we see Nike designers with hundreds of thousands of followers!

JOE FOSTER Societal shifts have seen an increase of influencers and available income to spend on sneakers. With this comes increased volume and opportunities for designers to think and act differently.

JULIANA SAGAT It's not about dropping a new sneaker on the market and doing a linear transaction anymore. It's about thinking of the life cycle of the product in a circular way, and considering longevity.

ALEXANDER TAYLOR Breaking rules, getting people out of the studio and into factories – that's what we've tried to bring to the work that we've done with adidas.

NIC GALWAY I'm drawn to the moments of change in brands. When someone had the courage to do something different, to take a different path.

ABOVE Vans vulcaniser production in the original USA factory, 1980

CHANGE

PEOPLE AND PROCESS

PEOPLE AND PROCESS

Designing a sneaker is a uniquely diverse process. As an industry, sports performance footwear has long been at the cutting edge of innovation, with everything from research and development to materials, manufacturing and, more recently, consumer data and insights contributing to its development. It is a world full of invention and big-picture thinking.

Many people play a role in the creation of a new sneaker. While representation within the sneaker industry hasn't always been balanced, companies today are making a conscious effort to move in a positive direction in terms of diversity among creatives. As our contributors attest, the more varied the voices involved in a process, the better the end result. Nothing good comes out of an echo chamber. Diversity comes in many forms, be it differences in race, gender, age, disability or education. It all leads to unique ways of identifying, understanding and solving problems, with each bringing individual value to the collective effort.

Throughout this chapter, we are offered a rare glimpse inside the minds of those involved in creating sneakers. Here we learn about their backgrounds and education, about their role within the creative process and what they have learned from working with others. Only then can we begin to understand the complexity, and strengths and flaws, of this industrial landscape. Discussions about how and where to seek inspiration, the fundamentals of good design, and how to define success, all help build a picture of the thought processes behind the products we take for granted every day as consumers.

In learning about the depths and layers of thought and design that go into creating a sneaker, they become so much more than just the functional objects they were made to be. Instead, they become physical representations of ideas, theories, dreams and accomplishments. They live on as timestamps of innovation and human achievement achieved through global collaboration.

PROCESS

SARA JARAMILLO There is no established, or right, or unique process. You learn to delete, add, edit and change your thoughts, processes and paths. You learn to listen.

STEVEN SMITH It starts with an intent: What is it for? Who is it for? How can I make it better? I am constantly gathering information and ideas from the world around me, which then become an archive to reference. Then I start sketching and ideating.

TUAN LE A designer translates sports requirements, footwear manufacturing know-how and current cultures, and makes them a reality.

CARLY MCKENZIE Each project is different. There's no set formula. Starting with a creative concept or narrative helps to connect the design brief to the consumer and authentically to the brand.

MURIEL JUNG I like to kick off a project with a very emotional, atmospheric approach before it leads to something tangible. I enjoy creating a cosmos around a product. I pin up mood boards, collect

ABOVE   Sara Jaramillo, prototype of ILYSM Tabi Sneaker, 2020

PROCESS

ABOVE   Asha Harper, prototype of Nike ISPA OverReact Sandal, 2018

materials, make very rough and expressive sketches and 3D mock-ups that describe the character, feeling and tonality of a product.

STEPHANIE HOWARD The first thing I do when designing any product is to make sure I understand the story it will be designed to tell. No matter if it is improving a performance metric or providing maximum cushioning, I want that story to show in the design language.

PETER FOGG My favourite part of shoe design was always sketching out ideas and solving problems in a new creative way.

SAMUEL PEARCE The early-sketch phase is my favourite part – when you have all those ideas in your head, and you start bringing them to life on the page. Don't be fooled, though: this is also the hardest part of designing. It's often as fun as it is tormenting, and you need time and a lot of patience to get those ideas out.

SARA JARAMILLO I love sketching, but my process doesn't start with a sketch, it starts with a real shoe – taking a vintage product, destroying it, and start building from that. For me, it's easier to start there and then sketch.

FRANCK BOISTEL After the brief has been defined, I take a few days to visualise the design in my head in 3D. I will then sketch, but I will not make multiple sketches. The benefit of visualising is that you see the design and build it in your head. Therefore, when you sketch it, it is ready, and the sketch becomes the 2D of your vision.

CHRIS SEVERN I conceive in my mind what I'm after, and then I define how I'm going to get there and what I'm trying to accomplish. In general, the way I operate is trial and error.

NIC GALWAY It's very important to take time to dream, to think things through. Only then is it really worth moving into realising and manufacturing.

ALEXANDER TAYLOR We don't design with 3D, CAD or Illustrator until we get to the point where we know it's going to work. You can fall into the trap where you've spent hours drawing something, and you don't want to change it, so you keep pushing it to make it work. We don't work like that.

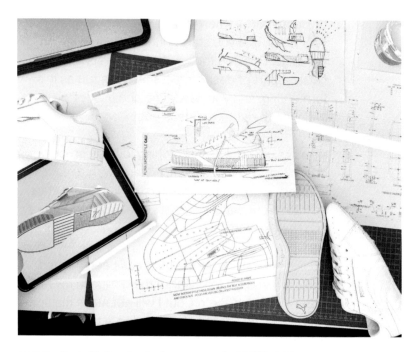

ABOVE Carly McKenzie, design process for PUMA, 2021

BENJAMIN GRENET I spend a lot of time with the athlete to listen and ask questions. After that, I try to translate, in a product, the mindset of the athlete. I only design when I want to capture all the elements I have from the athlete.

RIAN POZZEBON I like to meet with people, work with them, maybe create some sketches, maybe some ideas, but come back the next day. And I'll work through the night to deliver. I want them to feel excited that they sign off. You have them for that moment. Trying to chase them down a week later to get them to sign off on packaging or another design or whatever is difficult.

DAVID RAYSSE At Brandblack, the team is super small. I handle the footwear design and my partner and I collaborate on the look and feel of every shoe.

ABOVE   Tuan Le, design sketch of Reebok Alien Stomper, 1985

ABOVE   A-Cold-Wall* × Nike, Nike Air Zoom Vomero+ 5 'Redox'
being hand dyed, 2019

HELEN KIRKUM In my studio, I am the designer and the maker. I ideate the entire process, deconstruct all my raw materials, see the sneaker through to the final detail, box it and ship it.

SARA JARAMILLO In big sports brands, you have colour and material designers. In fashion, you're the one doing everything. I think about the colours from the beginning. I start designing with colours and materials.

PETER MOORE My approach to designing sneakers or anything is pretty much the same – I am more of a problem-solver than a designer. I discover the problem by talking to the athlete and finding out what their needs are and what they would like the shoe to do for them. From there, I talk to the innovation people to see what they might have to meet the athlete's needs. From there, I begin to design the shoe that might solve the problem. This is the ideal, and usually there are lots of other voices, like marketing, but they are often quieted by presenting the product as a solution to an athlete's needs.

SAMUEL ROSS I understand that I'm not a specialist footwear designer. When it comes to A-Cold-Wall*, it's more about communicating and conveying commentary and cultural touchpoints through footwear.

NICOLE MCLAUGHLIN My process has always been a fluid and organic approach rather than a forced one. When I started making footwear, it was about taking shapes that already existed and pushing them further.

JOE FOSTER   It is a long time since I was involved with designing sneakers, but my mission was to break into the US running market. I needed a five-star running shoe and succeeded with the Aztec.

ALEXANDER TAYLOR   My role has been until recently to develop ideas and deliver proofs of principles, whether it's a material, some textile that we've found, a particular machine, or a specific process. To put yourself in a place where you're looking at technologies from aerospace, automotive, or whatever it is. And to say, 'How can we plug that into making performance footwear?'

NIC GALWAY   My approach is to continually explore and make studies, and not worry too much if it's possible, but rather if it's worth figuring out. These studies often sit in the studio for some time and evolve until we see the right moment to connect them with a final brief.

STEVE MCDONALD   New design directions are incredibly difficult. Creating new ideas is painful but super fun, and satisfying when it works.

CARLY MCKENZIE   The most challenging project is always the 'follow-up design'. After a big hit, the challenge to recreate the success, or even maintain it, can be daunting! It is vital to understand that the brief must evolve along with the consumer, the trend and the brand to remain authentic.

CHRIS SEVERN   I lean toward designing things that don't go out of style, and the best examples of that would be the Superstar and the Stan Smith.

SARA JARAMILLO When YEEZY started, we were three people in the footwear and accessories team. adidas had a separate team, but Kanye wanted to keep his ideas, develop them and then give them to adidas. He wanted to keep a bit of control over the design process. He has a really different design approach – to mix fashion with sneakers and technology, without forgetting all those art and historical references.

MARINA CHEDEL As designers, we have to cover different areas. We have to be historians and study what has been done and how it was done. We have to be anthropologists, to dive deeper into our society to try to understand our consumers. And of course, we are creatives, generating new ideas and designing new products.

DAVID RAYSSE There's a balance between pushing against everything and trying something no one has seen, and being an egomaniac creating self-indulgent trash. I always try to remember that almost all of my favourite recording artists made shit music once they had complete control over every aspect of their world.

JACQUES CHASSAING Each designer has an ego like hell, and you always want to push technicians to go through with ideas. But sometimes, you have to find compromises.

CHRIS LAW There are so many ideas and samples that never make it to market. Getting a shoe from sketch to shelf is like an assault course, and there are many that fall along the way.

ABOVE adidas Primeknit samples by Alexander Taylor Studio, 2010

CHARLOTTE LEE As designers, we're in a constant moving state. We're always thinking of evolving our creativity and taking our learnings from the previous project to shape the next one.

STEVEN SMITH The best part is learning something new and then creating something that makes people go, 'Whoa, what is that?!'

CHRIS LAW My favourite part of the seasonal process is when you get to the factory. That's when it gets exciting. You learn so much in that time – it's like two years of college in two weeks.

TOM ASTRELLA Receiving that cardboard box from the factory containing your new samples is a feeling every footwear designer knows. It's like Christmas morning. All your hard work, all your sketches, have come to life and are inside that box.

ALEXANDER TAYLOR I wasn't ever interested in designing one shoe. I was interested in changing the way that shoes could be made.

SARA JARAMILLO We are historians. We tell our history in shoes instead of words. I would like to see our shoes in ten years and be able to learn how we got through a pandemic.

INSPIRATION

KIRSTEN SCHAMBRA Inspiration can come from anything at any time. How and when it comes out can be different for each project. The key is just to get started.

ANDREA NIETO I would like to change the perception that inspiration happens spontaneously. Generally, a good idea is the output of an involved process of research, curiosity, learning and collaboration.

FRANCK BOISTEL There is no end or limit to what you can draw inspiration from. It is all around us. Your eye is your first and best tool.

DAVID RAYSSE I try to sketch everything that pops in my head. It's important not to throw sketches away right away. Sometimes if I return to a sketch, I see something I missed that is intriguing and worth developing further. Ninety per cent of what I sketch is nothing more than a means to an end.

PETER FOGG Knowing what idea to pursue is not easy. I used to spend a lot of time sketching

ABOVE   Nicole McLaughlin, Reebok slipper, 2019
NEXT PAGE   Charlotte Lee, design sketch of New Balance 327, 2019

INSPIRATION

Asymmetric
midsole
flair

355 outsole lugs
wrap up heel

Fang toe
inspired by
355 / supercomp

over-sized
N logo
branding

ideas and options, so I could say, 'I explored the possibilities, and I think this is best.'

SAMUEL PEARCE As a designer, it's your job to know what looks good, and it usually hits you in a single moment. Between all the notes and scratchy lines of your sketches, something will usually stand out.

STEPHANIE HOWARD Story and technology guide my inspiration research. The best solutions weave the two together in an intuitive, unique way.

PETER MOORE My inspiration usually comes from either the athlete, as in the Air Jordan 1, or the brand, as in the adidas Equipment project.

JEAN KHALIFÉ I find myself more inspired when I interact with other creatives and share ideas.

HELEN KIRKUM I am mostly inspired by the things around me. By the ordinary processes of everyday life and how we interact with the products we own and love. I can see the memories embedded in the material, the wear and tear, the journey. That is the biggest inspiration for me.

STEVEN SMITH The motivation for me was seeing something that I wish I had designed. That is what pushed me to be better.

SARA JARAMILLO I work 24/7. I love what I do. Since it is an active and inactive process, you don't feel like you're working all the time, but you are. Being out and seeing people, analysing how we live, our purchases, how we walk, how other people walk, how we cross the streets – for designers, living is a job.

TUAN LE For the past forty years, I have discovered running as the true source of inspiration for whatever design challenges are thrown at me. It must be the adrenaline. I usually work out the complete design in my mind during my daily runs, without ever using a pencil. No matter what country or continent I travel to, a run every day is a requirement for my creativity and sanity.

PETER MOORE I think the best idea I have ever seen in footwear is the adidas Feet You Wear concept. It made it to market, but without much support. It created a unique look and feel when it was executed right ... which was not very often.

CHARLOTTE LEE The New Balance 327 was the sneaker that I had the most inspiration for. I had a really strong understanding of the brand's identity, and already had ideas in the bank that would refresh the archive in a way we hadn't explored before.

SAMUEL ROSS The NC.1 is probably my favourite sneaker of the past two years. There are clearly odes there to trail-running sneakers, the OG Oakley work, early 2000s Nike ACG, even some of the early Nike Air Presto and Free shoes.

STEVE MCDONALD The Nike Air Deschütz sandals, Nike Air Moc, Nike Air Azona, Nike Air Max Force, Nike Considered, Nike Considered Mowabb – every one of these was challenging. This is why my hair is all white.

ABOVE   A-Cold-Wall* NC.1, 2020

INSPIRATION

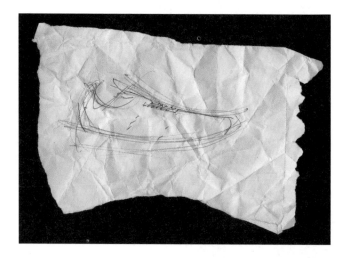

KIRSTEN SCHAMBRA At Nike, I remember struggling to come up with ideas for a shoe called the Silverfish – a lightweight training shoe for Olympic sprinters. I felt the pressure to create something amazing. I kept drawing blanks, kept trying to design, but nothing flowed. Then one night, I had a dream, and in it was the design. I was quick enough to wake up, roll over, find a scrap envelope on the floor and scribble a quick drawing down. When I woke up, I expected the sketch to be awful and throw it away, but to my surprise, the simple lines had potential.

SARA JARAMILLO The way Kanye pushed us to research and find new references was great. No one in the sneaker industry was doing that. He would tell us: 'I perform for hours. I jump on stage for hours. I'm an athlete, and there's no product for me out there because I'm not deemed an athlete.' At that time, Kanye was wearing the Ultraboost but the design wasn't meeting his aesthetic needs. So we offered an aesthetic answer to a performance shoe with YEEZY 350.

STEVEN SMITH Sometimes we have glimpses of the future that others can't see. The Reebok Fury was one of these projects. I had to stay true to the vision and keep the end goal in mind, even if it meant steamrolling opposition.

SAMUEL PEARCE Looking at what is in the market right now is a fatal mistake. If you're doing that, you're already two years behind everyone else.

BENJAMIN GRENET If you look too much at your own industry, you are just a follower, inspired by the last Nike or adidas or New Balance. I prefer to keep my distance.

ALEXANDER TAYLOR Knowing when to stop is really important. You find a lot of projects where there has been a huge investment to make something work because you're too far down the line to stop. And that's a problem.

OPPOSITE   Kirsten Schambra, design sketch of Nike Silverfish, 2001
ABOVE   Nike Silverfish, 2002

INSPIRATION

FUNDAMENTALS

JULIANA SAGAT When you design, the first goal is to create an innovative solution to make people's lives better and easier.

CARLY MCKENZIE Sometimes the simplest of ideas are the most successful. A small twist on a classic with the right timing can grab the attention of an entirely new audience.

BEN COTTRELL Familiarity and function are fundamental.

TUAN LE I've learned so much about not putting too much into a design, and not too little – just enough to make a perfect product.

STEPHANIE HOWARD Simplify where you can, amplify where it matters.

CHRIS HILL I always want whatever I'm working on to be better than the last design I made. That is a challenge in itself.

STEPHANIE HOWARD During the ideation process, it's important to go wide with options and not to home in

ABOVE   Nike Air-Sole cushioning units, 1992

early on one concept. Many sketches with different directions are explored. Curating those concepts is not always easy.

NIC GALWAY There is no such thing as work for the bin. I have ideas and samples we have made over the past twenty years that still inspire me, and when the time is right, they find their place.

MURIEL JUNG Many things are not feasible the way you imagined them. But it is always better to cut from a daring idea until it fits into reality, rather than build up from a safe and expected starting point.

DAVID RAYSSE Sometimes a design will come when it comes. You need to sit back and let it come to you. It will reveal itself.

CHRIS LAW The fundamental truths of sneaker design, when it comes to the actual product, are: character, tension and stance. As for the process, it's about finding a groove that feels right to you.

TUAN LE During my career, I've learned a few lessons that are still true today: be authentic, be technical, have restraint, be happy and healthy, and always be open to new ideas.

ALEXANDER TAYLOR A holistic working method allows you to spot opportunity.

JEAN-PHILIPPE LALONDE Colour is fundamental. One of Salomon's first steps in moving into new distribution was based purely on colour. In doing so, we introduced these shoes in the same exact shape to a totally different audience.

SAM HANDY There are a lot of performance fundamentals that don't change. What we're really doing is tuning at the edges. The human foot is the human foot.

MURIEL JUNG A design isn't credible if it isn't dedicated to its performance. Every detail needs to fulfil its reason to exist.

CHRIS LAW Do the best for the shoe and the consumer. You have to learn to take personal preference and bias out of most projects – it's

OPPOSITE   Pyer Moss × Reebok samples and prototypes, 2018

not very often you get briefed to make something you would wear.

TOM ASTRELLA The consumer doesn't always know what they want. No consumer would have asked for an airbag in their sole. But once presented with it, it turns out they liked it quite a lot.

STEVEN SMITH Make something you would want to pay for yourself. If you wouldn't buy it, why would someone else?

KIRSTEN SCHAMBRA The fundamental in any design is: who are you designing for and what do they need?

SARA JARAMILLO I believe small things make big changes. So maybe by wearing the same shoes, men and women are going to be treated the same in the future.

JULIANA SAGAT You need cultural context and insight. It's super important to be attuned to what's going on in the world in order to deliver a good design.

CARLY MCKENZIE Having a key insight gives you something to refer back to, and to test your designs against.

STEVE MCDONALD Trust yourself and your knowledge. Design it, build it, test it, repeat.

OPPOSITE   Salomon XT-6, AW19 campaign, 2019

FUNDAMENTALS

TOM ASTRELLA   Trust your gut, and the collective opinion of your peers. If everyone is saying, 'That looks cool', it's probably because it is!

NIC GALWAY   When everyone feels really good about something, then it's often too late, too familiar. I like an element of tension. We need to feel confident, but also a little uncomfortable.

JACQUES CHASSAING   Design should be purposeful. I like being disruptive too. You shouldn't stay too kind, because you have to wake up the consumer.

FRANCK BOISTEL   If it is original, if the design flows, if you are excited by a new design, then you know you have something worth pursuing.

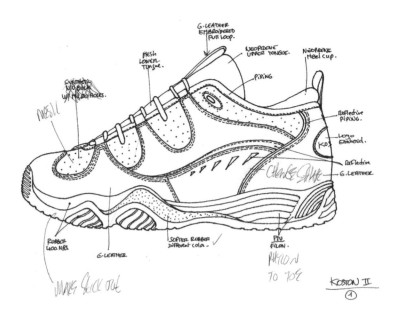

ABOVE   Franck Boistel, design sketch of éS Koston II, 1999

JEAN KHALIFÉ If you are in a position to challenge the status quo, you should get disruptive and push the boundaries.

SAMUEL ROSS Experimentation is a designer doing what they're supposed to do: liberating new ideas for society.

SAMUEL PEARCE Designers, by definition, have the power to force change. We are in a unique position to shape the future, and it's our responsibility to make a positive impact.

JEAN-PHILIPPE LALONDE The most important thing of all is that there's no such thing as a comfort zone. Expect the unexpected.

FRANCK BOISTEL Don't be afraid to be bold. Bold is beautiful!

ALEXANDER TAYLOR Everything can be improved.

RIAN POZZEBON You're designing for the future. You're always in the future, and the future is constantly evolving and changing.

FEEDBACK

STEPHANIE HOWARD A brand's goal is to create relationships with its consumers beyond just a monetary transaction. One way to achieve this is when insights are manifested in design solutions that make a difference in their lives.

PETER MOORE I don't really believe that design should be led by consumers. If design is led by consumers, what do you need designers for?

JACQUES CHASSAING The consumer should not lead the design, and the design should not lead the consumer. I think it's up to a designer to satisfy expectations.

ASHA HARPER There should be a give-and-take relationship between designers and who they are creating for.

SARA JARAMILLO I would love to say that consumers should be leading design. Because good design starts from a consumer need. It could be solving something aesthetic or something functional, but it should be solving something.

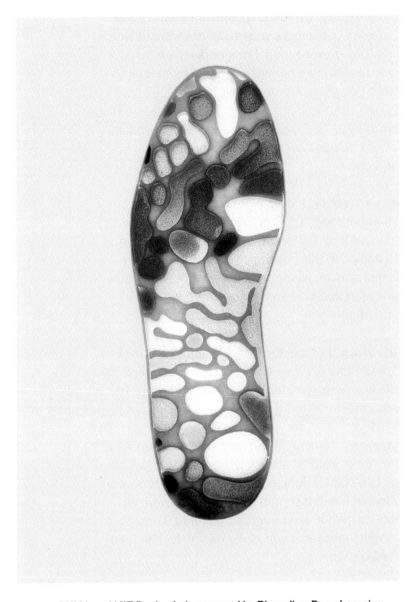

ABOVE   PUMA and MIT Design Lab, powered by Biorealize, Deep Learning
Insole, 2018. Using organisms, this insole measures long and short-term
chemical phenomena for real-time biofeedback on athlete fatigue
and well-being

CHARLOTTE LEE There is a constant conversation between consumers and designers without words ever being exchanged. Designers look to what consumers are wearing, and consumers look to what designers are designing.

NIC GALWAY We live in a time of direct feedback, where brands need to become more agile and open.

RIAN POZZEBON If it's a small niche of ratty skateboarders who are manipulating the shoe to meet their performance needs, you're going to design to this consumer. If something's on a broader level, you're going to want to push out and lead that consumer into something new and fresh.

TILL JAGLA For me, the magic moment is to see how people react when they see the outcome for the first time. Consumer sentiment is always the best, most unfiltered feedback.

STEPHANIE HOWARD Years before athleisure was a thing, our insights at Nike uncovered that our more mobile culture demanded a new kind of footwear. Women needed something that could work all day no matter what they had planned. These insights inspired my modular design for the Nike Air Max Verona.

DAVID RAYSSE I study how people react to products, not what they say. When someone likes your design, they won't put it down. This is what I look for.

SAM HANDY Talking to the person who has to use this thing as a tool and hearing their feedback and taking them seriously is important because their feedback is probably just as valuable as what your lab testing is telling you.

CHRIS SEVERN I worked through prototypes, and then the prototypes were tested in the best laboratory in the world: on the athlete.

RYAN FORSYTH A key factor in the process of making shoes is a constant flowing conversation. Having that consistent communication between all teams involved in the process – marketing, design and development – helps to create an amazing product.

ABOVE adidas Ultraboost 21 product testing, 2020

FEEDBACK

CHRIS HILL Feedback is critical to creating the best possible design. Bringing in other ideas and perspectives almost always plays a big part in that. If you can't humble yourself to self-reflect, self-critique, learn, be open to other perspectives and take a step back, you'll never get the best design. You can't be the best in anything by yourself.

NIC GALWAY Design critique is very important. It helps us unlock what we can't see alone.

FRANCK BOISTEL Feedback is essential, whether you receive it from your team, marketing, sales, management, or the professionals you design for. Without feedback, you are designing blind.

ABOVE   Nike Air Zoom Alphafly NEXT% prototype, 2018

CHARLOTTE LEE I see negative feedback as an opportunity to make something better.

RIAN POZZEBON When a product goes out into the market, I can't ditch the part where I didn't listen to someone's opinion. I have to own that part and learn from it.

JEAN KHALIFÉ The best feedback I ever received was to never be too emotionally involved in what I do. It is always a danger to fall in love with a design; you don't dare to change or challenge it enough.

NIC GALWAY The best advice would be from Yohji Yamamoto. When we first started working together, his key words to me were 'too noisy'. With that, he meant you have too many ideas, pick one with conviction. Twenty years later, I still remind myself of this every day.

SARA JARAMILLO The best feedback I've had is from Jeff Henderson: 'Make it simple. Even more. More. A little more.'

COLLABORATION

ASHA HARPER There is a diverse team of minds in and out of the design industry that help bring a vision to life. Like bees, we all cross-pollinate with each other to create magic.

CARLY MCKENZIE Sneaker design is a collaborative effort between the whole product engine: product management, design, development and marketing. Everyone plays an integral role in bringing a vision to life.

BEN COTTRELL From the initial idea and creation, to a product's final delivery, all team members, manufacturers, quality controllers and even the delivery driver play an essential role.

SAMUEL PEARCE The factory staff are really the unsung heroes in this industry. When you know the names of the people on the production line and the integral part they play in bringing each shoe to life, it really is a special thing to be a part of.

STEPHANIE HOWARD The two closest team members to the designer are the developer and the product

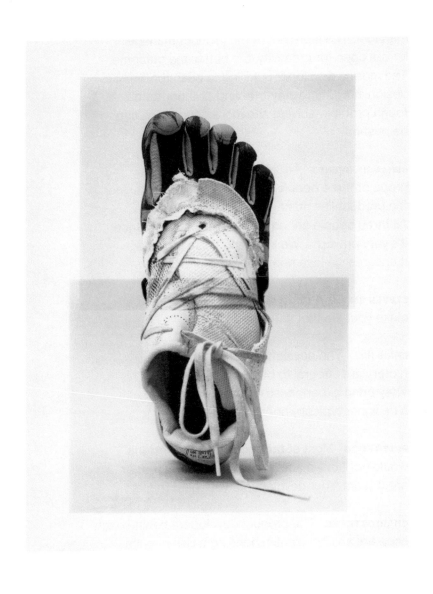

ABOVE Helen Kirkum × Matthew Needham collaboration, 2020

manager. The more doors the product manager leaves open for creativity, the better the outcome. The more a developer is willing to push for the design integrity to stay true and collaborate on tough problem-solving, the more boundaries can be pushed.

BENJAMIN GRENET At Salomon, there is a small group of three people for each range: the designer, the product line manager and the developer. All three people are equal in terms of importance. It's very powerful. We start and finish together. It ensures we lose nothing to interpretation.

STEVEN SMITH A good developer or engineer can make your life that much better.

CHRIS HILL The development team and the factory are integral to building a sneaker. They bring experiences and knowledge that a designer typically doesn't have.

PETER FOGG Most of the great sneaker designs would not have made it to the market without a great team. Your developer is your best friend.

CHARLOTTE LEE The product developers behind each shoe are so important in bringing a vision to life. Designers can create the most interesting concept, but it's the development team who get the concept to creation.

CHRIS LAW All your hard work needs to be matched with a developer who is understanding, has passion to fulfil your ambition, and brings their own additional input to improve the shoe.

MARINA CHEDEL My favourite part of the process is collaborating. Embracing different points of view and backgrounds will push the concept and the product further. It has the potential to make you discover things that you would have never thought about.

NIC GALWAY Working with collaborators, there is no set process. Each partner is different, and it's about building a relationship.

MATTHEW DAINTY Trust in your collaborators to use their skill set, ideas and technology to accommodate your vision. Play to their strengths.

CARLY MCKENZIE One of my first roles within PUMA was to work on the Hussein Chalayan collaboration. Working under his creative direction challenged me to unlearn my processes.

RYAN FORSYTH Partners always bring a fresh mindset to products, whether it's learning how to bring their brand story into a shoe, or looking at footwear from a completely new perspective.

TILL JAGLA When it comes to collaborations, it's vital to get a good understanding of the partner's DNA in order to create the best of both worlds. Just to slap a logo on a product ain't enough. The market is saturated. You need to find a smart, creative, but also intuitive way to create a product that is special enough to cut through the clutter.

ABOVE   Yongsoo Jeong walks in the **COTTWEILER SS18 show**, 2017

SAMUEL ROSS At the time of the Nike Air Zoom Vomero 5 collaboration, Nike London was really pushing the envelope on what London could do within the parameters of Nike. There was this open dialogue of experimentation. To be frank, I wanted to do more with that shoe. If I had more time to develop them, I would have made the whole mid-sole transparent with complementary tints. It would have been amazing. But this is where our dip-dyed iterations and in-house experiments came about, which led to the Solarized version. I was really keen just to keep that level of experimentation going.

JEAN KHALIFÉ It is important to embrace inclusion and diversity of opinion in order to push a project further.

CHRIS HILL Working with partners really opens your eyes to other ways of thinking and how other industries work.

HELEN KIRKUM When I started my studio in 2017, I wanted to explore how collaborations happen, to see if collaborators can redesign the product itself.

NIC GALWAY I've been very lucky in my career to work with many influential collaborators who I also see as mentors. The fundamentals I have learnt are: focus on one idea and make it memorable; have the courage to dream bigger; your compromise is someone else's starting point.

NEXT SPREAD **COTTWEILER for Reebok, Desert High, SS18, 2017**

COLLABORATION

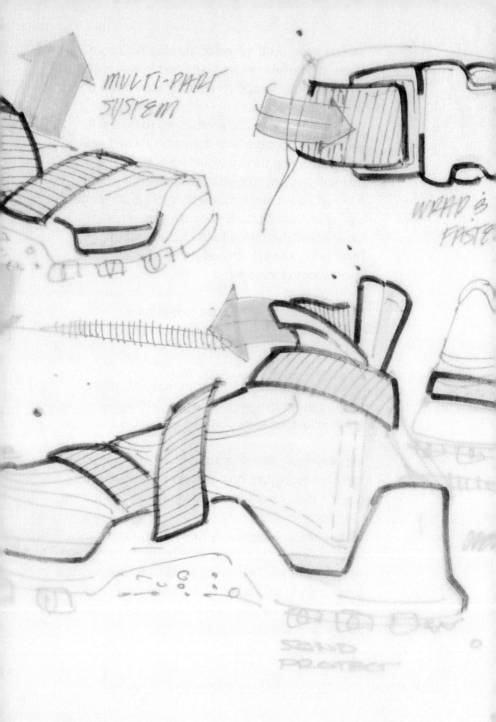

MULTI-PART
SYSTEM

WRAP &
FASTE...

S&ND
PROTECT

GAITER
FOLDS & PACKS
AWAY

DUAL TONG

COLLABORATION

ABOVE   Aria Polkey wearing black top and skirt by Comme des Garçons and Comme des Garçons × Salomon SS21 sneakers for *Purple* magazine, 2021

RIAN POZZEBON When you're approaching a collaboration, there's a conversation you have to have. Who is this shoe for? Your fans, or for you to personally wear?

ASHA HARPER The definition of collaboration is constantly changing, depending on whose court the ball lands in.

NIC GALWAY Working with Kanye when he first partnered with adidas really stretched me as a designer. He pushed me to look beyond the compromises we tend to make as creatives and to surround myself with the ideas of others.

SAMUEL ROSS When it comes to collaborations with sportswear partners, such as Nike or Converse, ideation and emotion lead the design process. What we produce will be merchandised on the annual calendar among a slew of other collaborators that they're working with, so it's really important that our product has a pronounced identity. I feel there's a responsibility to amplify the identity of the brand and where it comes from.

JEAN-PHILIPPE LALONDE Collaboration is a mirror for Salomon. Through these mirrors, we learn.

DAVID RAYSSE We are social creatures, and our industry is based on collaboration. How you see the world differently can only improve how I see the world.

TECHNOLOGY

NIC GALWAY Technology unlocks potential. It's at its most empowering when we get to explore how technology can shape a product rather than just as an ingredient.

PETER MOORE We are living through one of the periods most affected by design technology. Knit technology has changed the way shoes look, how they fit and how they perform. It is just the beginning.

JEAN KHALIFÉ Technology is good as long as it is used as a tool and not as something that will generate the design on its own.

TILL JAGLA It helps when technology makes a product better. But if you use technology just for the sake of it, the product will fail.

FRANCK BOISTEL Too much technology kills design, but technology in footwear is fundamental.

DAVID RAYSSE We live in a time where the craziest ideas I can think of can be more or less realised.

ABOVE   Eliud Kipchoge (KEN), white vest, wearing the Nike Air Zoom Alphafly NEXT%, and running behind a timing car projecting a laser grid which guided his pacemakers to keep an average of 2:50min/km splits. On 12 October 2019, he became the first athlete to run a marathon distance under two hours, in 1:59:40.2

The dawn of 3D printing and the revolutionary impact it is having on manufacturing is accelerating. I love it. Bring it on!

CHRIS SEVERN You can learn things by studying materials and putting them through stress and wear tests in the lab, but in the end, all the clever designing doesn't amount to anything if the athlete doesn't perform the way they want to in the product.

SARA JARAMILLO It's important to know when to use data and when not to use it. Sometimes you have your data, you release the product, and you think it is perfect, and it's not. Intuition also plays a big role.

CHRIS HILL Data is a huge part of the industry. It's important to a degree, but you can't rely solely on data to make your decisions.

PETER MOORE I hate data and the whole idea of data. It's another word for Big Brother.

STEVEN SMITH Sometimes it's good not to listen to market information because that represents now, and I choose to live in the future, not in what would be the past by the time the product hits the market.

PETER FOGG If you are on the leading edge of innovation and technology, you will have breakthroughs and success, but you will also experience some failure. Being the first to use the latest tech can lead to issues: it can make your shoe overpriced; problems can come up during wear-testing that require a rushed redesign; and the result may not be as good as was hoped for. We were going to use a new midsole foam compound on the Nike Hyperdunk 2014. The foam was adding cost, plus it wasn't as durable as it needed to be. In the end, we went back to the original foam. Luckily, the change didn't require a redesign!

JOE FOSTER The internet and other technologies have placed designers closer to their consumers, which can only be a benefit.

CHARLOTTE LEE Social media has enabled designers to be more aware, to learn about cultural moments, identify trends and see how consumers are reacting to product.

STEPHANIE HOWARD There are positives and negatives to social media and the attention culture it has created. If you can bypass the desire to grab attention and use social media for the great global connector that it is, then it is a beautiful tool.

JULIANA SAGAT Social media created this new pressure of staying fresh in every picture that you post. I miss the time when your parents would buy you a new pair of sneakers for school, and you would wear them all year long, and trash them, and be proud to have owned them.

ABOVE **KRAM/WEISSHAAR, Strung Robot, used to make adidas FUTURECRAFT.STRUNG, 2020**

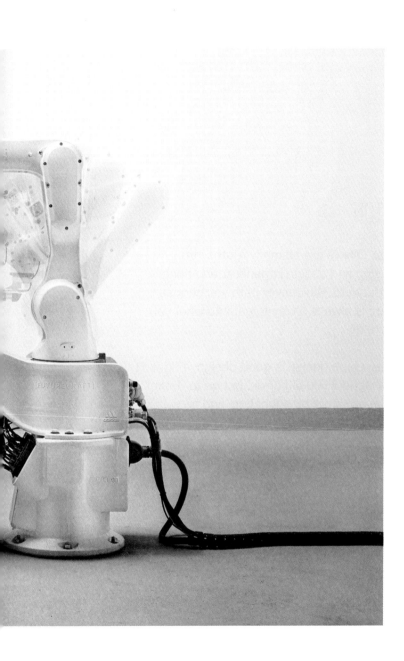

TECHNOLOGY

EDUCATION

HELEN KIRKUM There are so many routes into sneaker design and design in general, and none are straightforward. Whichever path you choose to take, there is always a way to get into what you really want to do.

PETER MOORE Good design is good design; it doesn't matter what school you did or did not go to. Tinker Hatfield is an architect by education, and he was a space planner for Nike when I found him and put him to work for me. Today, I think it is safe to say he is the world's foremost athletic footwear designer.

DAVID RAYSSE I'm old enough that there were no degrees in footwear when I started! Everything I learned about shoemaking I learned on the job. Whatever path you take can get you there if the will and work ethic are present.

NIC GALWAY Sneaker design has become much more established as a career opportunity. In this respect, it makes a lot of sense that there are opportunities to study it in its own right. But there

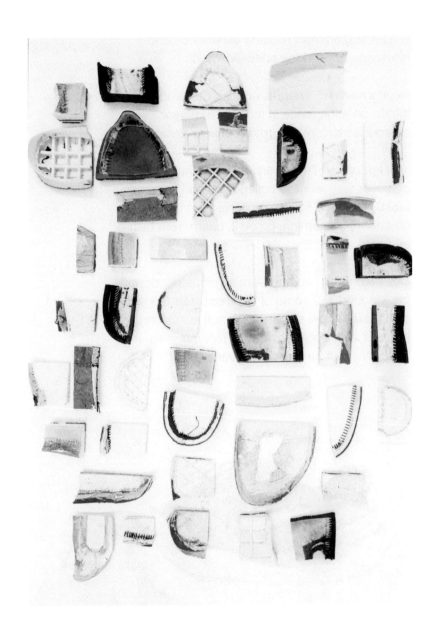

ABOVE   Helen Kirkum Sole Archeology, Soles, RCA Research, 2016

EDUCATION

is a lot to gain from building experience and reference beyond the sneaker world.

FRANCK BOISTEL There is nothing better than hands-on experience in multiple footwear categories to forge a great footwear designer.

RIAN POZZEBON I'm not a classically trained industrial engineer who went to college. I was influenced mostly by skate culture growing up in LA. To just go and try something, go do it. Learn from failure and try it again.

CARLY MCKENZIE I come from an industrial design background, but some of the most talented sneaker designers I know studied automotive, fashion, product and footwear design. This mix is what keeps the industry fresh and varied.

ABOVE Sara Jaramillo, sneaker experiment, 2020

SAM HANDY A lot of my experience came from working in retail. There's a lot to be said for having that kind of background and understanding of how the industry works.

HELEN KIRKUM My direct experience of studying and teaching shoemaking, technical footwear design and pattern-making allows me to experiment more freely.

CHRIS LAW I'm a true believer that formal education may help in some ways, but I don't see it as a must. What is a must is a real passion for footwear.

JULIANA SAGAT You have to develop a tolerance for failure. And I don't think that education teaches you that. If you want to be an innovator, this is super important.

JACQUES CHASSAING Creativity is not something you can learn in school.

STEVEN SMITH Natural skills and an artistic gift with clear thinking are far more important than a specific education.

JULIANA SAGAT Education can give you a good foundation, but if you want to be a creative in the modern world, then curiosity, flexibility and kindness are extremely important.

TOM ASTRELLA There is no singular path to becoming a footwear designer, just as there should be no singular way of conceptualising a sneaker.

PEOPLE

ASHA HARPER It's so important for industries to have a diverse collective mindset, breeding transparency, mindful conversations, constructive feedback, honesty, equality and equity.

DAVID RAYSSE When I started, there was very little diversity. What's great now is that it's not a bunch of white guys trying to guess what we want.

STEPHANIE HOWARD Fifty per cent of consumers are women, and, in commerce, women account for eighty-five per cent of all purchases and drive seventy to eighty per cent of all consumer spending. Yet most product decisions and designs are made by majority male teams. The inherent bias of those designs is not intentional, but it is real.

MURIEL JUNG Women-specific models in the performance range shouldn't be vastly different. Differences should be limited to small finishes and featural adjustments.

SAM HANDY I'm not sure we need to be making a more feminine running shoe for women, because

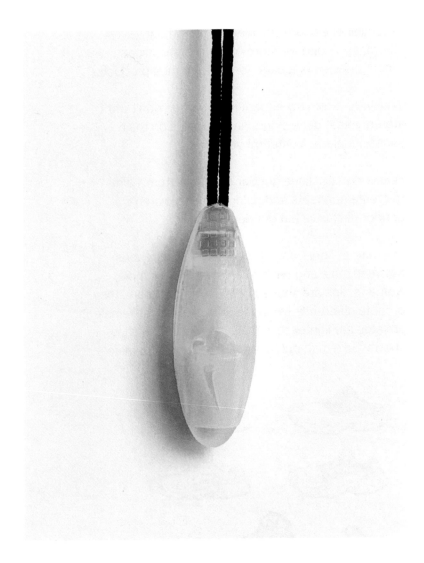

ABOVE Nike 'Run Safe' whistle, 2002. Designed by Stephanie Howard, the Air Max Verona and the Run Safe whistle were part of the first women-specific running footwear initiative from Nike, marking a shift to separate designs for women. Among other innovations led by new consumer insights, the shoes featured a secret pocket on the tongue that housed a safety whistle to wear on the wrist while running alone. A direct response to the unique vulnerability experienced by women runners

PEOPLE

its feminineness doesn't make it perform any differently. I hope for more of a universal unisex product design language for performance products.

CHRIS HILL From my experience, women don't want stereotypical 'girl' colours and graphics on their shoes, let alone a different version.

CHARLOTTE LEE I see gender on a spectrum. I aim to create concepts and colourways that aren't gender-exclusive but gender-inclusive.

SAM HANDY There are biomechanical differences between male and female athletes. For example, women's hips are shaped differently. Therefore, they pronate differently when they run. In general, female athletes are lighter than male athletes in the same sports. So a material that might deform for a male

ABOVE   Stephanie Howard, ideation sketches and final concept of Nike Air Max Verona, 2002

runner might be usable for a female runner and might deliver a different benefit. But in general, I believe the right way is to treat humans as humans, and running as running.

STEPHANIE HOWARD The design landscape should be reflective of the market. We know there is diversity in the market, and there is no reason why the makeup of the design departments should not reflect that.

CHRIS LAW Brands need to better mirror the communities and people that support them.

JULIANA SAGAT We still need to empower and uplift minorities. We need to see more diversity and create new role models. The sneaker industry shouldn't underestimate the power of diverse representations.

NIC GALWAY Diversity will be central to creating and building teams in the future. It is vital, as there has been a clear imbalance in the past.

STEPHANIE HOWARD It's the mix of ideas that come from diversity, equality and inclusion at the table that will create the future.

DAVID RAYSSE Diversity and the good it brings, different ideas and cultures – that's the spice of humanity. We need more voices, and we are getting them.

SUCCESS

CHRIS SEVERN I would define success as making a superior product. Sometimes it's major changes, and other times it's minor adjustments to an existing design.

PETER MOORE Success in sneaker design is simply: 'Does it solve the problem and does it fit the brand?'

FRANCK BOISTEL If your design becomes a classic and stays in the line for multiple seasons, then you have achieved success.

CARLY MCKENZIE I try not to define success by sales or even whether a design reaches the market. Sometimes the most inspirational designs never leave the studio but end up as the starting point for the aesthetic of a whole collection or category.

PETER FOGG Another way to look at success is with innovation and technology. If you can change the industry, change the market or sport with your footwear – that's exciting!

ABOVE   Nike Air VaporMax, 2020. The Flyknit upper is made with around sixty-seven per cent post-industrial recycled yarn and other recycled components, including plastic bottle waste. Flyknit was instrumental in ushering in a new era of knitted footwear, drastically reducing waste in manufacturing

SAMUEL ROSS Success in sneaker design is being able to push the conversation to a slightly more uncomfortable space and still attain a high level of sell-through. To move the purpose of footwear forward.

NIC GALWAY The success I value the most is how sneakers are viewed over time, their effect on culture and how they shape the future.

STEPHANIE HOWARD The biggest success in design is when you have made someone's life better and improved the process along the way. It's not just sales volume that defines success.

DAVID RAYSSE Does the work stand alone, devoid of hype and celebrity culture? Is it appreciated on design merits alone? That's success.

SAMUEL PEARCE Success has nothing to do with Instagram likes, or collaboration partners. Real design is about answering the brief and the continual improvement of the product.

JEAN KHALIFÉ It is a success if the product finds its audience. We are working in a very saturated industry. It is important to appreciate the moment when even a niche product gets a bit of attention.

STEVEN SMITH There is no better feeling than seeing your work out in the wild!

TILL JAGLA A successful sneaker design is something that adds value to a market that has seen almost everything already.

ASHA HARPER Success in sneaker design is about lowering your ego and having fun with freedom.

OPPOSITE   Undefeated × Nike Air Max 90, 2019

SUCCESS

INNOVATION AND THE FUTURE

INNOVATION AND THE FUTURE

The sneaker industry has always been built on
innovation. Over the decades, it has pushed
human achievement, helping to break records
and move manufacturing methods forward.
Adversely, it is also considered one of the most
damaging industries to the planet. A shift to
acknowledge social responsibility for this
is currently in motion, with some of the
world's best-known brands actively seeking
answers to the problems we face collectively
as human beings. In part, this is spurred on by
the rise of smaller, more nimble challenger
brands that are forging new paths. It is also
partly thanks to a shift in consumer demands.
Whatever the reasons, innovation engines within
giant corporations are being redirected.

   We have been here before. The 1990s saw a
move toward recycled and reclaimed materials
in footwear, spearheaded by innovations such
as the Reebok Telos boot and Nike Grind. The
2000s saw the introduction of lines like Nike
Considered, completely rethinking footwear
manufacture from design to production. More
recently, the age of knitted footwear ushered

in by Nike Flyknit and adidas Primeknit has drastically reduced waste at the manufacturing level by completely rethinking how sneakers are made, while outsider innovation such as Parley Ocean Plastic is continuing the important work of repurposing materials from waste. This is steady progress, but the recent increase in consumer consciousness for environmental responsibility is what will fuel an unprecedented momentum to improve sustainability within the industry.

A bold new future lies within reach. A future that promises shoes which don't steal from natural resources, and brands responsible for their product for as long as the raw materials within them exist. It's a vision where a pair of shoes is shredded when the consumer is done with them, and where that material is in turn transformed into the next pair. At its most ambitious, the industry is aiming for individualised shoes that are printed at home, or grown using mushroom mycelium. The idea is to focus on sustainability, and in doing so, vastly reduce the carbon footprint of the sneaker industry without compromising on style, design, comfort and performance.

For now – as many of our contributors confirm – we should be wary of calling anything 'sustainable', perhaps opting instead for 'responsible'. While work on sustainability is more prevalent than ever before, there is much still to be achieved. Our sustainable utopia may be a way off, but at last it seems that people at all levels see it as our future.

PROGRESS

ROMAIN GIRARD The way shoes are built today is still very archaic. The main portion is made by hand. It is very exciting to be in a moment where a lot of things are changing. Access to new processes, technologies and materials allows us to completely rethink how we build shoes.

DAVID RAYSSE We are at the end of the mass-produced era. We are at the dawn of customised designs, manufactured in vending machines right in your local warehouse and shipped to you in hours. It's gonna be like streaming shoes. It will radically reduce waste and carbon footprints as we will forgo shipping, warehousing and trucking, and avoid the huge number of shoes that end up in landfill.

TUAN LE Most of the industry is built on cheap labour in Asia. In the 1980s, it was South Korea, then Taiwan, China, Thailand, Indonesia, Vietnam. We have left behind so many trash heaps along the way. Manufacturing has to change. There are a few companies now that build their products out of garbage – foam made out of leftover foam, rubber outsoles made from old rubber outsoles.

ABOVE   ILYSM Metal Tabi, 2021

PROGRESS

BENJAMIN GRENET When you change a production tool, you also change the way to design the shoe. It's a permanent back and forth between what we want to design and how we will produce it. And our next challenge is to reinvent our production tools.

FRANCK BOISTEL There's a 'beat the competition' mentality in footwear, and every company wants to be the first to come up with this or that. It pushes us to be innovative as designers.

JACQUES CHASSAING One concept I did in the mid-1990s with Peter Moore was Feet You Wear. When we were doing tests, having a guy with basketball shoes jumping and landing, the main problem was ankle sprain. Doing the same thing barefoot, it didn't happen. So we asked ourselves: 'How can we reflect the anatomy, motion and structure of a foot in a shoe?' The foot has smooth edges and is soft. So we introduced this kind of curve and the softness into a shoe.

PETER MOORE I don't believe the industry is moving forward in a positive way. It is being driven by big numbers, not by innovation.

ROMAIN GIRARD I had the chance to be part of the first spark of the knitting revolution in footwear, working with Alexander Taylor. What felt nearly impossible back then is standard today.

ALEXANDER TAYLOR Not much innovation had happened since Nike Air. That's why there was an impetus at adidas to anticipate change and make it happen.

STEVEN SMITH I really liked being the co-champion of algae-based materials added to the plastics we use. Bringing manufacturing back to the US has also been a very satisfying aspect of my role.

SAM HANDY NMD was a really interesting shoe from adidas. It was the first time that we took top-end performance technology, and the influence of the archive, and pulled them together. We put Boost in NMD while Boost was still the newest performance ingredient. That did something brand new.

JEAN KHALIFÉ If we think about experience for the consumer, Boost remains a good standard that went from peak performance benefit to something that everyone adopted casually. The fact that we are starting to recycle it with our FUTUTRECRAFT.LOOP initiative is an innovation on its own.

ABOVE   Sara Jaramillo, design sketch of ILYSM Tabi tooling, 2019

SAMUEL ROSS The Air Force 1 lows we released with Nike used Flyleather, produced in Peterborough. It's an engineered leather made with at least fifty per cent recycled leather fibre. The success of the A-Cold-Wall* Air Force 1 was almost used to propagate the material innovation and push it into a commercial stance.

SAM HANDY In five years, we went from presenting one shoe with Parley at the UN to every single pair of our biggest shoe – Ultraboost – containing recycled sea plastic. It's mad! But it's only a beginning.

JULIANA SAGAT At Nike, we are looking to 'Move to Zero', our journey towards zero carbon and zero waste. It's a long journey, but I'm really optimistic.

ABOVE   Alexander Taylor, adidas FUTURECRAFT.Tailored Fibre prototype, 2015

SAMUEL PEARCE 3D printing has yet to really catch on due to its high price and material limitations. Once that has caught up to industry standards, we will see a shift in the landscape like never before.

NIC GALWAY The blurring of the physical and digital worlds we live in today will unlock different thinking and opportunity.

DAVID RAYSSE People need to appreciate quality over hype. I'm not sure we are there yet as social media is accelerating the hype phenomenon.

JEAN KHALIFÉ The main positive moving forward is to refocus on function rather than style. Sustainability is also currently a key element that is very important for the years to come.

SAM HANDY Health and well-being – that's what's going to influence the footwear industry the most. Sport is starting to mean something different to people. Look at the effects of the COVID-19 pandemic: people are thinking about running differently, what it does for them and their mental space.

TOM ASTRELLA The pandemic has pushed everyone to think differently. Previously, footwear component suppliers relied on trade shows to sell their products. Footwear designers were regular visitors to factories in Asia. All of that has changed. We are currently collaborating with software companies, exploring digital remote working, replacing what we would normally undertake in person. These new developments and this type of working have the potential to change things forever.

ABOVE   FKA twigs × Nike, FKA twigs and Miles Chamley-Watson in the do you believe in more? campaign, wearing Free Transform Flyknit and Metcon DXS Flyknit, 2017

ALEXANDER TAYLOR I can imagine the pandemic is unwittingly speeding up the process for people to start making from home or find new ways to make – as a cottage industry. That next wave will come, and you've got to solve all the problems that come with it. There's so much to solve. But that's the creative juice. As a designer, you've got to be thinking like that.

JULIANA SAGAT We are at the beginning of a big shift. Climate change and human rights are in the spotlight. The sneaker industry needs to tackle these issues in order to create a better future.

SARA JARAMILLO Right now, we as an industry are creating needs. That's the reality. I would love to think that design can go back to the simple things – solving functional needs.

ASHA HARPER We have to reinvent what success means in sneaker design when the planet and humanity are prioritised. A lot of the challenges we face are because our values are not aligned with our intentions.

SAMUEL ROSS I'm really interested in the conversation about selling the intellectual property of a brand to the consumer in a physical format. It's where Service Point 1 – a small project that we launched last year – is still kicking on. The idea of being able to de-rig the costs associated with a brand logo and give that to an educated and informed consumer to see how they incorporate that into their own design or application. You still allow people to purchase, but the ticket of entry is far lower, which is way more democratic.

SARA JARAMILLO The customer is asking for gender fluidity, and the planet is asking for more sustainability. Those are the main problems right now.

ASHA HARPER Slow progress is better than no progress, but even that is a privilege when people aren't equally impacted.

JULIANA SAGAT I'm so optimistic about the future of this industry. There is a new wave of creative women willing to change the status quo from the inside.

MURIEL JUNG What if we didn't have to limit ourselves in the consumption of upcoming fashion trends because resources constantly regrew and waste seamlessly merged back into nature? That would be amazing.

INNOVATION

SUSI PROUDMAN Innovation in manufacturing is critical in the design process of footwear. We're continually evaluating what we would like to make with how we can make it.

NICOLE MCLAUGHLIN Innovation is most impactful when it's about helping people and the planet. We need to look past products and focus on everything that goes into them, from manufacturing and packaging to durability and community.

STEPHANIE HOWARD Many companies don't invest enough in their innovation engine, including patience, not just cash. Nike invests with patience and you see the rewards of that through their sustained dominance in the market.

CHRIS HILL My favourite innovation over the last few years is hands-free footwear. Obviously, it's great because it helps people, but it's also truly innovative.

ALEXANDER TAYLOR adidas FUTURECRAFT.Tailored Fibre used a process that came from the world of technical embroidery. It was used to make heated

ABOVE   PUMA, BioEvolution, 2019. This project uses biologically active
materials to adapt uniquely to the wearer's foot

car seats and parts of aeroplanes and rotary blades on military helicopters. My motivation was to find something that we could build into textiles that would give stability and fusion and replace moulded parts. The first prototypes only just hang together, but they proved an idea.

DAVID RAYSSE Knitting, injection-moulding, 3D design and 3D printing have significantly changed how we manufacture shoes. We have so few limitations now.

ALEXANDER TAYLOR I had exposure to working with what we referred to at the time as 3D knitting. It was about rationalising that whole flat-knit process and saying, 'could we have an upper essentially popping off one machine?' And that was when I proposed researching and working with flat-knit technology to adidas, which later became Primeknit.

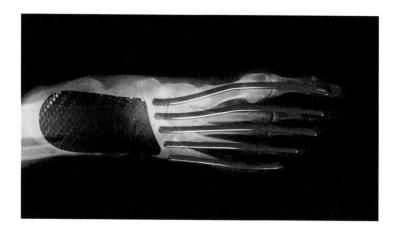

ABOVE adidas, adizero adios Pro, 2020. X-ray reveals adidas EnergyRods – five tuned carbon-infused rods, which mimic the metatarsal bones of the foot
NEXT SPREAD KRAM/WEISSHAAR, Strung Robot, creating adidas FUTURECRAFT.STRUNG, 2020

JACQUES CHASSAING I'm excited about 3D printing – having this one-piece shoe coming out of a printer. It's innovation not only from a creative point of view, but it also reduces manufacturing and material costs.

SAM HANDY 3D printed performance – it's wild. The idea that you can individually tune every facet of a running midsole to work together in a kind of kinetic chain. It's a fascinating intersection of design and engineering, with tools that none of us have ever used before.

ROMAIN GIRARD We are seeing unexplored spaces when it comes to manufacturing innovation. For a couple of years now, we have been exploring biodesign – looking at how living organisms can help to build material, bond material, shape products and so on.

ANDREA NIETO adidas FUTURECRAFT.STRUNG breaks traditional footwear constraints because using this, we can place yarns in any direction. We're not bound by the traditional warp and weft. It allows us to create a more dynamic textile that matches the way the human body is made.

SAM HANDY I'm really excited about adizero and EnergyRods. We reproduce the metatarsal bones of the human foot with enhanced materials underneath the foot itself. There's something a bit special about the idea of biomimicry – enhancing the human body with materials that are not found in the human body.

ALEXANDER TAYLOR We're finding new ways to make footwear here in the studio, as a kind of incubation thing through the pandemic. We've brought tools and machines in here that ordinarily we wouldn't have done. And we've found that working next to the machines compresses the whole process, fast-tracking everything.

NIC GALWAY The industry is looking at innovation through a very different lens than in the past, investing much more deeply in sustainability, both in the products we build but also in the impact we have through our processes. There is collective energy to accelerate learning and to make change.

SUSI PROUDMAN We're seeing a lot of interesting innovation in leather-type products, with cowhides replaced by apples or plants, replicating the same texture but without the carbon footprint or chemicals associated with traditional leather.

RYAN FORSYTH Reebok has had a lot of interesting eco-technology that flies under the radar, like Cotton + Corn, where you can essentially bury your shoe in the ground, and it will return to the natural elements in the earth over time.

BENJAMIN GRENET Salomon INDEX.01 is our first totally recyclable running shoe, made with a new foam material. The engineer who worked on the foam wasn't happy with the marbled effect. We discovered it in the bin, and we said, 'Oh, man, this is so nice!' With our eyes as designers, we saw this 'imperfect' look as a strong visual signature.

TUAN LE At Mizuno Japan, I am part of a team working to create 3D printed shoes right in front of a consumer. Next to the kiosk, there will be a place for consumers to drop off old, worn-out shoes. We are planning to recycle those old shoes into other products.

SAM HANDY FUTURECRAFT.LOOP is an exciting innovation. You shred the shoe, and you get the raw material back, and all for the environmental cost of turning on a shredder.

SARA JARAMILLO I'm really excited about FUTURECRAFT.LOOP by adidas. It could solve the sustainability problem. Let's see!

ABOVE   Salomon INDEX.01 recyclable running shoe, 2021

SUSTAINABILITY

SAMUEL PEARCE There are no quick solutions for what's happening in the world right now, but we have to start reversing our impact.

SARA JARAMILLO I would encourage the industry not to talk about sustainability because it is lying to the customer. When they sell you a sustainable product, you think it is the greenest thing ever. And it's misleading. They should be telling you it is a more responsible product.

PETER MOORE The critical driver for sneaker design is the effect it has on the environment. The shoe you design today could be on the planet for at least 100 years.

TOM ASTRELLA Being one of the worst polluting industries in the world is not a title that any industry wants. So the future of footwear must be about becoming more sustainable.

BENJAMIN GRENET Sustainability is what we need to be driven by today. We have to totally change the game. We are in an experimental phase at Salomon.

ABOVE   Nike Cortez with Flyleather, an engineered leather made with at least fifty per cent recycled leather fibre, 2018

MURIEL JUNG There's a huge opportunity for worldwide companies to have a significant impact contributing to a less wasteful industry. And that journey starts with the designer's work. A product not only fulfils consumer needs, it also shapes trends for the future.

PETER MOORE Designers will need to become more demanding in the materials they use. Methods like knitting are effective, but they should demand that this technique advances and innovates.

HELEN KIRKUM The industry has to work towards creating more sustainable products. To categorically change its approach to overproduction and consumption. There has to be drastic and immediate action to create headway.

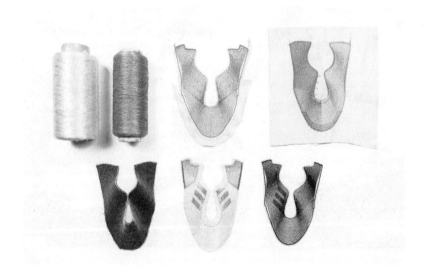

ABOVE adidas' Parley upper development made with Parley Ocean Plastic®, 2019

ALEXANDER TAYLOR We can work as hard as we possibly can, but it needs to be a global effort if it's really going to have an impact. Having an influence on what's going on in different parts of the world is very difficult, but you need everyone on board to make a big difference.

SAM HANDY The industry needs to completely change how it makes things. Capitalism has given us a pretty unhealthy way of treating the world, and we need to find a way to fix that. We need to keep making things for people, but in a way that doesn't break the world.

JOE FOSTER I think we are some way from changing consumer behaviour. Working towards recycling sneakers is a better direction for designers.

SAMUEL ROSS Some of the conversations I've been having with Nike revolve around the idea of taking away options from collaborators and pushing sustainable materials by default. At this point, I don't feel that should be a choice. It should be a default.

KIRSTEN SCHAMBRA I continue to explore eco-friendly materials and manufacturing. I want to build brands that design with that in mind from the get-go. I'm currently testing vegan cactus and grape skin 'leathers', and leather made from the byproducts of the leather industry.

SUSI PROUDMAN Sneakers are a highly coveted, highly traded product, but they're also predominantly made of plastic. If we make products more naturally and sustainably, and reduce our carbon footprint while keeping the design, the hype, the fun – then the legacy that we're leaving to the generation after us isn't a pile of plastic shoes in a landfill. It's a story of real change.

JACQUES CHASSAING I hope we will have no more plastic bottles. It's nice to say, 'I'm cleaning up the beaches or cleaning up the oceans,' but you have to solve the problem at the beginning and avoid having plastic waste in the oceans or on the beaches.

KIRSTEN SCHAMBRA If we can create products by using waste from other industries, and when the life cycle for this new product ends, it can in turn be upcycled into something else for another industry, then the idea of a big-industry upcycling collaboration can become a reality.

TUAN LE The change in the manufacturing of shoes from trash will dramatically change the look of the sneaker industry. The next big challenge will be to create a complete cycle of products, from waste to new product, again and again.

JULIANA SAGAT Brands should be accountable for their design forever. When the consumer can't wear the shoe anymore, they should be able to give it back, and the waste derived from it should be utilised to create a new design.

STEPHANIE HOWARD Many brands are working towards a closed-loop, circular product cycle. Some are experimenting with natural, biodegradable and even bio-fabricated products. Some are aiming at the next level up – regenerative products that leave

ABOVE   Nike Space Hippie, 2020. An exploratory footwear collection constructed with Nike's own 'space junk', transforming scrap material from factory floors into a radical expression of circular design. Each shoe is made with at least twenty-five per cent recycled content by weight the earth better than they found it. It is an exciting time to be involved in product design. For too many decades the basic tenets of making something good were ignored.

ASHA HARPER If it can't be reduced, reused, repaired, rebuilt, recycled or composted, it should be restricted, redesigned or removed from production.

MURIEL JUNG The dream would be to create a product that is in harmony with nature. Bio-based and biodegradable materials are just the beginning.

DAVID RAYSSE We need to stop thinking of sneakers as disposable. Stop throwing away your shoes. Make better products that can live on.

CHARLOTTE LEE Wearing really good-quality products that can be repaired or loved in an aged state massively reduces consumption. It changes how we desire and buy products.

HELEN KIRKUM Design has a massive role to play in showcasing and encouraging the longevity of products. I try to encourage that mindset within my work. I promote the beauty and joy of wearing, and I reinvigorate treasured possessions.

STEPHANIE HOWARD There is a lot of waste at the factory level that people never see. Designers can and should be allowed to innovate to improve this. Those leaps are risky, and there is no guarantee of return on investment, but they are necessary.

MATTHEW DAINTY The production of high-quality products will ultimately result in less waste.

JEAN-PHILIPPE LALONDE If you're producing something sustainable at the other end of the world to cater to the European or North American market, it kind of defeats the purpose of implementing this technology.

ASHA HARPER We should study the effects of consumer consumption and challenge how much we produce and why – and what the long-term effects of these choices could mean.

SAMUEL ROSS A huge part of the future of material and production is ensuring, when we move into using more sustainable, safe, forward-thinking, futuristic materials, that there is complete visibility on how every person within that process is actually being enabled. It's not just about the end product.

ALEXANDER TAYLOR The best thing that big brands can do – which they are doing – is to create incubation hubs and start-ups within the brand that build sustainable models.

FUTURE

FRANCK BOISTEL There will be an explosion of new innovations and creativity when the COVID-19 crisis is over. Footwear people can't travel as before, factories have been hurt, the market is changing. I can't predict what will happen in a few years, but I am sure we will see new technologies, new shapes and patterns. Exciting footwear will emerge from this.

CHRIS HILL Hopefully, it will swing back around so that the best design wins again. Or maybe the best of both worlds will combine: hype is based on great design first and foremost, and then who wears it.

MURIEL JUNG I'm excited to see how designer's tools will evolve. There are already VR design programs that allow you to create with goggles and controllers in your hands. But this gestural creation is just the beginning of what augmented reality could mean to the design process.

PETER FOGG I see a lot more computer work in the future! I think designing and sketching in 3D would be great. It has to become as easy as the 2D digital drawing apps.

ABOVE   Roe Ethridge, Mehdi on a Motorcycle, 2019. Dye sublimation print on aluminium, 60 × 40 in (152.4 × 101.6 cm), image courtesy of the artist and Andrew Kreps Gallery, New York

FUTURE

SAM HANDY The idea of human needs and sport being a bit more intrinsically connected will make us make slightly different products – probably more democratic products.

CHRIS SEVERN Injury prevention or reduction will be a very valuable contribution to the world of sports shoes. That's what I see in the future.

STEPHANIE HOWARD The future of sneaker design is more of the good and less of the bad. Designers care about the future. They want to create the most amazing products.

NIC GALWAY One area I am very interested in is how we can evolve manufacturing, and how that can help shape the future aesthetic of what we create.

TOM ASTRELLA Labour is the most expensive factor in footwear production. It was the reason almost all sneaker production moved to Asia. But even there, the cost is increasing. The future is automation to reduce cost, and localisation to reduce carbon footprints.

STEPHANIE HOWARD There will be more automation at the factory level and more intimacy at the consumer interaction level.

STEVEN SMITH Local manufacturing and sources of materials are the critical drivers for the future. It addresses all the great ideas. Growing eco-materials on site and assembling them as close as possible to the consumer.

SAM HANDY Materials are going to need to change, but to change them, you have to change what people find valuable. We need to redefine what value looks like. And that's going to be a slow process. What people are used to thinking of as valuable is plastic and shiny.

NIC GALWAY I see the future as less about small increments in human ability, although this will always be important, but rather using the resource we have to further sustainable thinking.

STEPHANIE HOWARD In the future, designers will utilise their knowledge of cultural shifts and build greener, more comfortable, adaptable, customisable, higher performing, durable footwear. The constant influx of new technologies will both overwhelm the process as well as move it forward at a rapid pace.

CHARLOTTE LEE We will see a massive shift in consumption in the coming years, from consumers wanting better quality, to how we manufacture products. With an increased focus on quality over quantity, manufacturing will need to adapt.

SAMUEL PEARCE Personally, I'd like to see a return to 'buying better for longer'. The idea that what you invest in becomes a more long-term purchase that can be repaired, or its components replaced to extend the product's life cycle.

HELEN KIRKUM We have to stop the overproduction of sneakers and be selective about what we create. We have to be driven by humanity and empathy. Sneaker design on a mass scale can no longer be solely

ABOVE   Nike Adapt BB lace engine, 2019. The Adapt BB marked the long-awaited realisation of Nike's dream for self-lacing footwear that began with the futuristic vision of the Nike MAG in the film *Back to the Future Part II*

FUTURE

driven by aesthetic and newness. It has to address the wider world. Its existence cannot only be for profit.

MARINA CHEDEL   I wish projects would be design-led and not business-led. It would mean that we would understand the power that creativity could have. We would continuously design with the best intentions for humans and our planet, or even our universe.

ASHA HARPER   We must replace 'egosystems' with 'ecosystems'.

HELEN KIRKUM   The future of sneaker design is personal. Sneakers show people who we are: they are our self-expression; they have to be loved and be worn.

NIC GALWAY   The future of sneaker design will continue to evolve. We have the opportunity and responsibility to invite more diverse voices into the industry to help shape its future.

ASHA HARPER   If we dared to deconstruct our restraints rather than being overly obsessed with our individuality, we'd realise that altruism has always been our future.

OPPOSITE   adidas, FUTURECRAFT 4D, 2017. The world's first high performance footwear featuring midsoles crafted with light and oxygen using Digital Light Synthesis, a technology pioneered by Carbon, a Silicon Valley-based tech company working to revolutionise product creation through hardware, software and molecular science

FUTURE

ADVICE

STEVEN SMITH Know your history, but don't be ruled by it. Do what is right, not what is easy.

ASHA HARPER Designers shape the world and are responsible for the world they shape. Be conscious of what you put out and how you think it will impact the world.

JACQUES CHASSAING Be open-minded. Be curious. Be responsive and be proactive. Rather than being reactive, dare to be irreverent, disruptive.

JEAN KHALIFÉ Please don't be too nerdy about shoes. You won't innovate much if it is the only thing you know in life.

CHRIS HILL Always be open to learning and getting better. There will always be someone better than you – so stay humble and open-minded.

NIC GALWAY Use every day to be creative, to be curious, and to learn something from someone.

ABOVE   Ancuta Sarca, personal archive, 2020. Sarca is an advocate for circular design, reworking kitten heels with old sneakers as an eco-friendly alternative

ADVICE

HELEN KIRKUM Build a great network and cherish their experience and opinions.

ALEXANDER TAYLOR There's nothing that can't be achieved if you believe in something enough. Just don't try to do it alone. You can never have enough gifted collaborators.

PETER MOORE Get involved – don't just sit there and wait for an order. Get involved with the marketing people and the production people. Become a pain in the butt.

STEVE MCDONALD A lateral side shoe design is about one per cent of the journey of your product. The rest of making it happen is ninety-nine per cent of it. Just make stuff and make mistakes and ask why it works or doesn't work.

STEPHANIE HOWARD It's not enough to have a great idea. You have to keep pushing that idea forward because the big ideas that really have an impact are complex. Recognise complexity and build a strategy to reach goals along the way.

STEVEN SMITH If someone tells you something can't be done, prove to them that it can!

FRANCK BOISTEL Never give up, never doubt yourself. Keep your eyes open, your mind clear and your curiosity intact.

ASHA HARPER Document everything; carry a journal.

STEVE MCDONALD Everything means everything. That is my best inspiration. Try not to pigeonhole ideas. Try to look at multiple perspectives at the same time. This is not easy, but try it.

BEN COTTRELL Experience the product first-hand to confirm that what you are designing is thought through.

STEPHANIE HOWARD Hard work is a necessity, and patience is essential for big leaps in innovation. It's easier to keep doing things the same way, but big rewards never come with little effort.

ALEXANDER TAYLOR Be patient and remember that if you really stick to something and it's good, and you believe in it and back yourself, then it will end up somewhere. If you make something good, people will find it.

JEAN KHALIFÉ Be attentive to what is around you, be passionate, and try to master the design codes and craft.

STEPHANIE HOWARD Never stop reading and learning. It's not enough to have one great skill set. We need to master many skills in order to solve the complex challenges we face together.

HELEN KIRKUM Try to understand your worth and your point of view, and allow that to be your power.

JULIANA SAGAT Nurture your difference because this is what makes you unique.

ASHA HARPER Actively seek out discomfort and embrace its value in harnessing awareness and humility into how we design.

FRANCK BOISTEL Go outside, live, play, travel, meet people, visit all the shoe stores you can, touch shoes, feel them, come back, draw or write. Create from your travels, not from browsing the internet. It is not going to make you a good designer.

SAMUEL ROSS Consider what footwear looks like across three to four different generations. There's so much scope for amazing footwear to be speaking to Boomers, Gen X, Gen Y and Millennials simultaneously. True innovation has the ability to percolate designs across society as a whole. Think as wide and as big as possible about that, versus serving a niche.

ABOVE   Tinker Hatfield, design sketch Nike Air Jordan III, 1987

SAM HANDY Take what has been done to date and run with it because it will be a completely different industry in five or ten years. Right now, the changes haven't really happened. The changes have only started.

STEPHANIE HOWARD Focus on what is important and don't be distracted by shiny new trends.

CHRIS SEVERN Shoot for something that is not only timely but also timeless.

JACQUES CHASSAING Do something to make the world better. Do something to make people better. Do something to make life better.

ASHA HARPER Things that excite you are not random. They are connected to your purpose.

ROMAIN GIRARD Seek a collaborative approach. Follow your guts. Create with passion. Never give up.

NIC GALWAY Always remember what drew you to sneaker design. Never lose sight of that.

STEPHANIE HOWARD Our calling is to build a bridge to the future, and celebrate the steps along the way.

ADVICE

MANIFESTO

THE SOLUTION TO THE PROBLEM IS INHERENT IN THE PROBLEM ITSELF.
PETER MOORE

DREAM OF A BETTER FUTURE THROUGH BETTER DESIGNS. NOT NEW AND DIFFERENT, BUT NEW AND BETTER.
STEVEN SMITH

DESIGN, NOT FOR YOURSELF, BUT FOR THE PURPOSE OF SOMEONE ELSE.
BENJAMIN GRENET

IS IT NECESSARY? ARE YOU CREATING SOMETHING THAT DOESN'T NEED TO BE CREATED?
BEN COTTRELL

TAKE RESPONSIBILITY AND BE ACCOUNTABLE FOR WHAT WE DO AND PUT OUT IN THE UNIVERSE.
MARINA CHEDEL

MAKE IT SUSTAINABLE.
ANDREA NIETO

SEEK OUT AND WORK WITH ONLY ECO-FRIENDLY MATERIALS AND PROCESSES AND LET THOSE CONSTRAINTS DICTATE DESIGNS AND CREATIONS.

KIRSTEN SCHAMBRA

BUILD WITH ZERO WASTE AND ZERO CARBON FOOTPRINT.

TOM ASTRELLA

CONSIDER THE EARTH.

JULIANA SAGAT

CREATE A SAFE SPACE FOR EVERYONE TO BE THEIR TRUE SELF.

MARINA CHEDEL

IF IT GOES AGAINST YOUR VALUES, SPEAK UP. ACCOUNTABILITY IS A MUST.

ASHA HARPER

MAKE ALL PRODUCT CUSTOM-FITTED FOR EACH AND EVERY INDIVIDUAL.

TOM ASTRELLA

PAVE THE PATH FOR THE NEXT GENERATION, KEEP THE DOOR OPEN, BE INCLUSIVE.

JULIANA SAGAT

LEARN FROM EACH OTHER AND FEED EACH OTHER AND NEVER FORGET TO HAVE FUN.

MARINA CHEDEL

PLANT SEEDS EVERY DAY.

ASHA HARPER

LEARN TO UNLEARN — TO RELEARN A NEW WAY.

JULIANA SAGAT

MORE ART. MORE AUTHENTICITY.

SARA JARAMILLO

NO MORE EGO.

JULIANA SAGAT

GO PLAY!

STEVE MCDONALD

MANIFESTO

BIOGRAPHIES

TOM ASTRELLA is co-founder of footwear design consultancy The Footsoldiers. His work includes the Under Armour Sports Style range (Forge), New Balance Reengineered and R_C franchises, and PUMA × McQueen line.

FRANCK BOISTEL is a footwear designer. During his time as Design Director for Sole Technology, Franck created an iconic lineage of skate shoes for éS, etnies and Emerica. His work includes the éS Scheme, Penny Pro, Viero and Muska Pro, etnies Calli-Cut and Emerica Reynolds.

JACQUES CHASSAING is a creative direction consultant. His work at adidas as footwear designer in the 1980s gave birth to the adidas Basketball Shoe Forum and the Running ZX franchise, repositioning both adidas and the running category as a whole. In the 1990s, Jacques helped rewrite performance footwear design again with adidas EQT and Feet You Wear.

MARINA CHEDEL is a footwear designer at Nike, having previously worked at Vivobarefoot. A London College of Fashion alumna with a master's in Footwear Design,

Marina works on big picture conceptual concepts. Her work includes the Air Force 1 NDSTRKT.

BEN COTTRELL AND MATTHEW DAINTY are the founders of fashion house COTTWEILER. Their work has helped redefine the relationship between sportswear and fashion. COTTWEILER's ongoing partnership with Reebok continues to push the boundaries of manufacturing and materials.

PETER FOGG is a retired footwear designer previously working at Nike. Peter created many iconic Nike outdoor sneakers that saw pop-culture success in the late 1990s, including the Nike Air Humara, Air Terra Humara, Zoom Air Terra Sertig and Air Terra Albis.

RYAN FORSYTH is Global Product Manager for collaborations at Reebok. His work includes partnerships with COTTWEILER, Vetements, Maison Margiela, KANGHYUK and Mountain Research.

JOE FOSTER is the founder and former CEO of Reebok. His family has been instrumental in developing running footwear since JW Foster & Sons was founded in 1895. Joe went on to found Reebok with his brother in 1958, where his work includes the Reebok World 10 and Aztec.

NIC GALWAY is Senior Vice President of Global Design at adidas Originals. Having worked with collaborators including Yohji Yamamoto and Kanye West, his work includes the Y-3 Qasa and YEEZY 350. Nic has overseen the renewal of historical adidas franchises, including Tubular and Equipment, and created the bestselling NMD line.

ROMAIN GIRARD is Senior Head of Innovation at PUMA, having entered the world of footwear design through PUMA in the early 2000s. He previously worked at adidas for almost ten years, helping to introduce Primeknit. His work at PUMA continues to push the boundaries of footwear innovation, including PUMA Jamming, XETIC and BioEvolution.

BENJAMIN GRENET is Footwear Concept Design Manager at Salomon. His work includes the XT-Wings, S/Lab Sense and XT-6, among many others.

SAM HANDY is Vice President of Design for running at adidas, having previously worked in the Originals and football departments. His work includes the adidas Originals ZX Flux, adidas Ultraboost, 4D Run, Solarboost 3 and adizero Adios Pro.

ASHA HARPER is Creative Lead Designer at Nike. Asha previously worked as an engineer and designer at Dr Martens and PUMA. Her work includes the Rihanna × PUMA FENTY Creeper, Nike ISPA OverReact Sandal and Jean Paul Gautier × sacai × Nike Vaporwaffle Mid.

CHRIS HILL is Senior Design Manager for energy collaborations at Reebok, having previously worked at adidas, Nike and The North Face. His work has brought together Reebok's world with Amber Rose, Cam'ron, Future, Kendrick Lamar, Pyer Moss and even Minions.

STEPHANIE HOWARD provides design and innovation consulting to global brands, having previously worked at New Balance, Nike and Reebok between 1994 and 2006. Of her many designs, notable sneakers include the New Balance 850, 825 and 875, Reebok Trail DMX6, and one of the first women-specific running footwear design initiatives at Nike, producing the Air Max Verona, Air Imara and Air Max Aware.

TILL JAGLA is Global Head of Energy at adidas. His work uses storytelling and collaborations to amplify key franchises for the brand, including partnerships with A Bathing Ape, Lego, Overkill and Sean Wotherspoon.

SARA JARAMILLO is a footwear designer and consultant. Having started in fashion, working at Proenza Schouler, she was one of the first designers to join the newly formed YEEZY. Working alongside Kanye West, she helped create the YEEZY 350. Her more recent work includes new footwear sneaker brand ILYSM and their Tabi Sneaker.

MURIEL JUNG is a footwear designer at adidas. Her work has spanned the football, outdoor and running departments, and includes the adidas Ultraboost 21.

JEAN KHALIFÉ is Footwear Design Director of Advanced Concepts at adidas. His work includes the adidas Originals NMD_R2, Atric F/22, adidas D Rose 10, and Zone Boost.

HELEN KIRKUM is a footwear designer and artist. Her work dismantles discarded sneakers to rebuild entirely new one-of-one silhouettes. In turn, these sneakers challenge our preconceptions of a sneaker's life cycle and our relationship with footwear in general.

JEAN-PHILIPPE LALONDE is Fashion Program Manager at Salomon. Since joining the brand, Jean-Philippe's work has brought a new interest in reviving the Salomon archive, including the XT-6, XT-Wings 2 and Odyssey.

CHRIS LAW is a multi-disciplinary designer with over fifteen years of experience in the footwear industry. His journey started at online forum Crooked Tongues before moving into senior design director positions for adidas Originals, Converse and Clarks Originals.

TUAN LE aka the 'Zen Master of Footwear Design' is a design consultant. His work over the last thirty years has given us the Reebok Freestyle and Alien Stomper (for the *Aliens* movie), AND1 Tai Chi, Merrell Moab and Mizuno Wave Rider, among many others. He continues to work closely with clients, including Mizuno, to push their innovation and design.

CHARLOTTE LEE is a footwear designer at New Balance. Her 327 design helped introduce a new approach to reinterpreting archive franchises at New Balance.

STEVE MCDONALD is a retired footwear designer and innovator, having previously worked at Fila, Nike, Under Armour and Apple Computer (frogdesign). His time at Nike saw him start the ACG franchise and co-create the Considered programme. Steve championed early innovations such as no-sew manufacturing and segmented midsoles (Nike Free). His work includes the Nike ACG Deschütz Sandals, Air Moc, Free Trail 5.0, ACG Escape, ACG Air Azona and Kobe 3, and Fila Scapegoat.

CARLY MCKENZIE is Head of Design for Sportstyle footwear at PUMA. In her role, she leads the creation of the Inline lifestyle collection. Her work continues PUMA's long history of challenging footwear lacing norms, including the PUMA Basket Heart.

NICOLE MCLAUGHLIN is a concept designer and consultant. Her work repurposes objects and materials to create new functional footwear. In doing so, she invites conversation around the future of sustainability, product design and what footwear can be.

PETER MOORE is the former (and first) Global Creative Director of Nike, before taking on the same role at adidas. His work is instrumental to the success of both companies, and continues to impact the sneaker industry today. Peter is responsible for the Nike Air Jordan franchise, adidas Equipment franchise, adidas Feet You Wear franchise, and for creating the Air Jordan 'Jumpman' logo and adidas Performance logo.

ANDREA NIETO is an innovation designer at adidas. Her work looks as far as seven years in the future and includes innovations such as adidas FUTURECRAFT.STRUNG.

SAMUEL PEARCE is Creative Design Manager at New Balance. His work often references his personal sneaker archive, hiding 'Easter eggs' in his sneaker designs, which include the New Balance X-90, X-Racer and Made in England line.

RIAN POZZEBON is Senior Director of Global Footwear Design at Vans. Rian is responsible for the Vans Vault, Vans California and Classics lines. He co-created Vans Syndicate, making sneakers with WTAPS, Shawn Stussy, Ice-T and Tyler, The Creator.

SUSI PROUDMAN is Chief Product Officer at the sustainable footwear brand Allbirds. She runs her own consultancy, Rose and William Consulting, where she helps start-ups with disruptive, sustainable mindsets to thrive. Previously, Susi spent seven years as Vice President of Materials at Nike across multiple departments, developing and driving end-to-end materials strategies, processes and tools.

DAVID RAYSSE has been working in fashion and athletic footwear design for over twenty-five years. He is the co-founder of Brandblack. David began his career in 1993 at Fila where he designed the Grant Hill II and later the Stackhouse. In 1997 he moved to adidas and designed shoes for Kobe Bryant and Antoine Walker.

SAMUEL ROSS is founder of A-COLD-WALL*, and has helped redefine streetwear, repositioning it for the modern age. Ross has collaborated with Nike, Converse and Dr Martens. The A-COLD-WALL* × Nike Air Force 1 helped launch Flyleather. The multiple iterations of the A-COLD-WALL* × Nike Air Zoom Vomero+ 5 pushed collaboration and customisation boundaries.

JULIANA SAGAT is a footwear designer at Nike. Before Nike, Juliana worked at Marc Jacobs, Kenzo, Givenchy and Isabel Marant. Her recent work for Nike has refreshed the Air Max line, including the Air Max 90 Recraft, Air VaporMax 360 and READYMADE × Nike Blazer.

KIRSTEN SCHAMBRA is a design and innovation consultant. Between 1999 and 2005 she was a footwear designer at Nike, creating the Nike Air Max Ltd, Tuned Air 7, Terra Switch, Silverfish and City Knife II.

CHRIS SEVERN is a self-titled 'Creationeer' based in California. Between 1956 and 1993 he has been involved with adidas US, France and Germany in a variety of ways, but his first love is creating new products or improving existing ones. Chris has recently re-connected with adidas US in Portland, Oregon as an independent consultant. His work includes the Superstar and Stan Smith.

STEVEN SMITH aka 'The Godfather of Dad Shoes' is a footwear designer who has worked at adidas, Fila, New Balance, Nike and Reebok. He is currently Design Director at YEEZY. Steven's work includes the New Balance 996 and 1500, adidas Artillery, Reebok Instapump Fury, Nike Air Zoom Spiridon Cage, Zoom Streak Spectrum and Air Max 2009, YEEZY 700 and YEEZY 450.

ALEXANDER TAYLOR is an industrial design consultant. Working with adidas, he has redefined how sneakers are made, co-creating Primeknit, FUTURECRAFT.Tailored Fibre, FUTURECRAFT Leather and the original Parley shoe.

DANIEL TAYLOR is Design Lead for Sportstyle footwear at PUMA. His work has reimagined archival PUMA franchises, including the RS-X, Future Rider, Nitefox and Tsugi Disc.

INDEX

INDEX

PICTURE CREDITS

Every reasonable attempt has been made to identify owners of copyright. Errors and omissions notified to the publisher will be corrected in subsequent editions.

Cover image by Jack Harper

PICTURE CREDITS

ACKNOWLEDGEMENTS

This book is based on the exhibition *Sneakers Unboxed: Studio to Street* at the Design Museum, London, 18 May to 24 October 2021.

Exhibition Curators
Ligaya Salazar and Shasti Lowton

Assistant Curator
Rachel Hajek

Curatorial Research Volunteers
Naomi Zaragoza and Suha Hwang

Fact Checker and Interpretation Consultant
Thomas Turner

Interpretation Editor
Maria Blyzinsky

Exhibition Project Managers
Nicola Underwood and Rebecca Gremmo

Exhibition Coordinator
Maximilian Kennedy

Exhibition Design
Interesting Projects Ltd.

Graphic Design
Studio LP

Lighting Design
Beam Lighting Design

Exhibition Build
Central Leisure Developments

The Design Museum would like to thank all lenders
and artists who have generously contributed to the
exhibition and the wider events programme.

Acquisitions made with the generous support of the
Conran Foundation.

With special thanks to:
Salman Ahmed, Ben Assefa-Folivi, Wayne Berkowitz,
Brett Booth, Theodore Chambers, Beatrix Ong,
Pensole Academy, Alex Powis, Justin Ronné,
Statista, Alex Williams

Exhibition Sponsor

Design Museum Publishing
Design Museum Enterprises Ltd
224-238 Kensington High Street
London W8 6AG
United Kingdom

designmuseum.org

First published in 2021
© 2021 Design Museum
Publishing

ISBN 978-1-872005-53-9

Publishing Manager
Mark Cortes Favis

Editor
Alex Powis

Consulting Editors
Ligaya Salazar and
Morgan Weekes

Assistant Editor
Giulia Morale

Picture Editor
Michael Radford

Copyeditor
Rebeka Cohen

Proofreader
Simon Coppock

Designer
Chris Benfield

Based on the exhibition design by
Studio LP

Many colleagues at the
Design Museum have
supported this book,
and thanks go to them all.

Distribution

UK, Europe and select territories
around the world
Thames & Hudson
181A High Holborn
London WC1V 7QX
United Kingdom
thamesandhudson.com

USA and Canada
ARTBOOK | D.A.P.
75 Broad Street, Suite 630
New York, NY 10004
United States of America
www.artbook.com

Printed and bound in Italy by
GRAPHICOM